Material characterization tests for objects of art and archaeology

Second edition

Nancy Odegaard
Scott Carroll
Werner S. Zimmt

Chemical equations by David Spurgeon
Illustrations by Stacey K. Lane

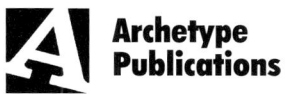
Archetype Publications

The authors and publisher can take no responsibility for any harm or damage to collections that may be caused by the use or misuse of any information contained herein.

The health and safety information contained within this book is provided for guidance only. It does not, and cannot, replace a thorough assessment of the health and safety risks associated with the use of hazardous materials and how to control them.

Laws exist in many countries to control the use of hazardous materials and readers are advised to contact the health and safety authorities within their own countries for further information.

The Threshold Limit Values (Occupational Exposure Limits) quoted were correct at the time of printing. However, these limits are regularly reviewed and readers are advised to seek out the latest information from either the manufacturer/supplier or their health and safety authority.

First published 2000 by
Archetype Publications Ltd
6, Fitzroy Square
London W1P 6DX

Second edition 2005

© Archetype Publications, 2005.

ISBN 1 904982 09 3

All rights reserved. No part of this publication may be reproduced, stored in a retrieval system, or transmitted, in any form or by any means, electronic, mechanical, photocopying, recording or otherwise, without the prior permission of the copyright holders.

Designed and typeset by Kate Williams, Abergavenny, UK.
Printed and bound by Gutenberg Press Limited, Malta.

Contents

Foreword for UK readers   *vii*

Acknowledgements   *xi*

1 Introduction   **1**

2 Chemical safety   **7**

3 Scientific method and techniques of spot testing   **19**

4 Spot tests for metals   **33**

    Test for aluminum ions using aluminon (aurintricarboxylic acid)   34

    Test for aluminum metal using Alizarin Red S and electrolysis   36

    Test for antimony using spot-test papers   38

    Test for arsenic compounds using spot-test papers   40

    Test for arsenic or phosphorus in copper metal using iron(III) chloride   42

    Test for chromium using diphenylcarbazide and electrolysis   44

    Test for copper using spot-test papers   46

    Test for copper using nitric acid and ammonia   48

    Test for copper using potassium ferrocyanide   50

    Test for copper using rubeanic acid   52

    Test to determine gold karat using a touchstone   54

Test to determine gold quality using nitric acid	56
Test for gold using tin(II) chloride and electrolysis	58
Test for gold plating using aqua regia	60
Test for iron using potassium ferrocyanide	62
Test for iron using hydrochloric acid	64
Test for lead using spot-test papers	66
Test for lead using potassium iodide and electrolysis	68
Test for lead using potassium dichromate	70
Test for mercury salts using diphenylcarbazone	72
Test for mercury using aqua regia	74
Test for mercury-cinnabar using a melting point apparatus	76
Test for nickel using spot-test papers	78
Test for nickel using dimethylglyoxime	80
Test for silver using potassium dichromate	82
Test for silver using potassium dichromate and electrolysis	84
Test to determine silver quality using nitric acid	86
Test for tin using cacotheline using electrolysis	88
Test for tin using sulfurous acid	90
Test for zinc using ammonium mercuric thiocyanate	92
Test for zinc in copper alloy using electrolysis	94
Test for zinc using spot-test papers	96

5 Spot tests for inorganic and ionic materials — 99

Test for calcium using nitric acid and sulfuric acid	100
Test for carbonate using hydrochloric acid and barium hydroxide	102
Test for chloride ions using sulfuric acid	104
Test for halogens (chlorine) using pyrolysis (Beilstein test)	106
Test for chloride using silver nitrate	108
Test for chlorine in polymers using pH paper and pyrolysis	110
Test for nitrate using spot-test papers	112
Test for nitrate using iron(II) sulfate	114
Test for phosphate using spot-test papers	116
Test for phosphate using ammonium molybdate and benzidine	118
Test for phosphate using ammonium molybdate and ascorbic acid	120
Test for sulfate using spot-test papers	122
Test for sulfate using barium chloride	124

6 Spot tests for organic materials — 127

Test for starch using iodine/potassium iodide	128
Test for simple carbohydrates using o-toluidine	130
Test for complex carbohydrates using o-toluidine	132
Test for carbohydrates using triphenyltetrazolium chloride	134
Test for triglycerides using triglyceride reagent	136

Test for unsaturated oils using potassium permanganate	138
Test for blood using benzidine	140
Test for protein (nitrogen) using calcium oxide and pyrolysis	142
Test for protein using copper(II) sulfate (Biuret test)	144
Test for sulfur using lead acetate paper and pyrolysis	146
Test for organic sulfur using calcium oxalate and pyrolysis	148
Test for phenols in vegetable-tanned leather using lead acetate	150
Test to determine vegetable-tanned leather using iron(III) sulfate	152
Test for indigo using sodium hydrogen sulfite	154
Test for lignin using phloroglucinol	156
Test for rosin using sulfuric acid (Raspail test)	158
Test for cellulose using aniline acetate and pyrolysis	160
Test for cellulose and its derivatives using 1-naphthol (Molisch test)	162
Test for nitrate (cellulose nitrate) using diphenylamine	164
Test for poly(vinyl alcohol) using iodine/potassium iodide	166
Test for polyester groups using hydroxylamine hydrochloride	168
Test for polyamides using p-dimethylaminobenzaldehyde	170
Test for polycarbonates using p-dimethylaminobenzaldehyde	172
Test for silicone-based rubber using pyrolysis	174

7 Other tests — 177

Test of pH using an Insta-Check® pH pencil	178
Test for acidity with pH pens	180
Test for volatile acids with pH papers and lime water	182
Test for volatile acids with pH papers and glycerol	184
Test for acidic vapors using cresol red	186
Test for hardness with a pencil sequence	188
Test to determine specific gravity using an electronic analytical balance	190
Test for radioactivity using photographic film	192

Appendix 1: Dilution table and chemical concentration calculations	195
Appendix 2: Explanation of pH	201
Appendix 3: List of abbreviations	205
Appendix 4: Table of reagents with safety information used in spot tests	207
Appendix 5: Chemicals, equipment and supplies	211
Appendix 6: Product suppliers	215
Appendix 7: Materials Characterization Trial Form	217
Appendix 8: Glossary	221
Appendix 9: Bibliography	227

Foreword for UK and EU readers

In the UK the use of hazardous materials is regulated by the Control of Substances Hazardous to Health Regulations 1994 (CoSHH) and subsequent amendments. The Health and Safety at Work etc. Act 1974 also places a general duty on all employers, and the self-employed, to ensure their own safety and the safety of others. Breaches of either piece of legislation may result in criminal prosecutions.

The CoSHH regulations implemented a European Union Directive on the safe use of hazardous materials. Similar legislation exists in other EU countries.

Substances hazardous to health

These are defined as:

- substances marked:
  Very Toxic, Toxic, Corrosive, Harmful, Irritant (see Chapter 2 for pictograms);
- substances with a maximum exposure limit (MEL) as specified in schedule 1 of CoSHH, or an occupational exposure standard (OES) – MELs and OESs are the UK equivalent of the threshold limit value in the USA;
- biological agents capable of causing an infection, allergy, toxicity or other human health hazards;
- any dust at a substantial concentration in air;

- any substance not listed above that creates a health hazard comparable to any of those listed.

Asbestos, lead and materials hazardous because of their flammability are covered by other legislation.

General requirements of CoSHH

Assessment of health risks

Before work with hazardous substances can take place, a suitable and sufficient assessment of the risks created by that work must be undertaken. The purposes of the assessment are: to identify what health and safety risks are presented to those exposed to the substance(s); to identify how those risks will be controlled; and to demonstrate to the health and safety enforcing authorities that the regulations have been complied with. The assessment must be written down for future reference and made known to those exposed to the substance(s).

The assessment should take into account such things as:

- what is being used (hazard data sheets or materials data sheets are available from suppliers and manufacturers of hazardous substances);
- how, where and for how long the substance is being used;
- the quantities involved; and
- the individuals involved (pregnant and nursing mothers are prohibited from using certain substances under UK and European legislation).

The assessment must be kept under review and, should there be any changes in the way the substance is to be used, the assessment must be repeated.

Prevention or control of exposure to hazardous substances

Once you have established that a substance is hazardous and that it presents a risk to the health and safety of those exposed to it, the risk must be eliminated or controlled.

Exposure to substances with MELs is required to be kept to as low a level as reasonably practicable. This can usually only be done with local exhaust ventilation (LEV). A MEL is not a target to work to, but the maximum level of exposure allowed.

Substances with OES are different, in that the legal duty is to keep exposure at or below the standard set.

Whatever the substance, the CoSHH regulations set out how exposure to it should be controlled. Called the "hierarchy of control", it sets out that risks to health and safety must be controlled at the source wherever possible, that the the use of personal protective equipment (PPE) to control exposure should only be considered as the last resort.

The hierarchy of control requires risk exposure to be controlled in the following order of priority:

- do not use the substance
- use a safer alternative substance
- install LEV or another engineering method of control
- provide PPE on its own or in conjunction with another means of control.

If engineering methods such as LEV are used to control exposure they must be regularly checked and maintained by a competent person. For example, LEV systems must be checked at least every 14 months. The regulations set out how often control systems must be checked. Records of the maintenance checks must be kept.

If PPE has to be used, then it must comply with the Personal Protective Equipment at Work Regulations 1992. These require: the PPE to be appropriate for the job and the user and manufactured to approved standards; the user to be trained in their use; and a regime of checking, testing and maintaining them to be in place. A record of PPE checks should be kept.

Monitoring exposure to hazardous substances

The CoSHH regulations require exposure to hazardous substances to be monitored to ensure that users are not being overexposed and that the control methods are working. For some substances, the regulations specify what is to be tested and how often. Monitoring regimes require specialist equipment and so expert advice may be necessary. Readers are advised to seek further information.

Health surveillance

Schedule 5 of the CoSHH regulations requires those exposed to certain substances used in certain ways to be regularly checked for adverse health effects. In addition, the regulations require employers to implement a system of regular health checks if this would add to the protection of the exposed workers through the early detection of adverse health effects.

Information, instruction and training

Users of hazardous substances must be provided with such information, instruction and training as is suitable and sufficient for them to know the health risks associated with their exposure to the substances with which they work. They must also be aware of the measures taken to control those risks and how to operate them.

Users should have free access to the risk assessments and all the other records required to be kept under the regulations. This information should be kept in a comprehensible format and should take into account those for whom English is not their first language.

Those involved in carrying out hazardous substances assessments must be competent to do so and may require specialist training.

Further reading

L5, *Control of Substances Hazardous to Health, Control of Carcinogenic Substances and Biological Agents: Approved Code of Practice*. Health and Safety Executive (HSE).
EH/40 (revised annually), *Occupational Exposure Limits*. HSE.
HS(G) 37 *Introduction to local exhaust ventilation*. HSE.
HS(G)97 *A Step-by-step Guide to CoSHH Assessment*. HSE.

Priced and free publications are available from the HSE. Consult a telephone directory for your nearest HSE office.

The HSE enforces health and safety legislation in the UK. In some circumstances this role is delegated to the Environmental Health Officers from the local authority. Both sources are prepared to give information over the telephone, but they will not undertake the assessments for you.

Acknowledgements

In compiling this study, we have enjoyed the advice and collaboration of many colleagues. We would like to acknowledge the support and assistance of many individuals who have brought the materials characterization project to publication.

The participation of Gretchen Voeks, National Park Service, Western Archaeological and Conservation Center during the early phases of research is greatly appreciated for the contributions of information, the acquisition of metal coupons, and beta testing of numerous tests. Numerous colleagues including Brigid Sullivan, Jim Roberts, Dr Colin Pearson, Helen Ganiaris, Holly Anderson, Dr Frank Katterman, Ellen McCrady, Dr H. P. Schramm, Lyndsie Selwyn, Pam Hatchfield, Helen Coxon and Dr Chris Tahk offered noteworthy encouragement and sent references, manuscripts and notes for possible inclusion in the volume. Likewise, the audience comments from presentations made to the American Institute for Conservation in 1996, the Western Association for Art Conservation in 1997, and the Canadian Association of Conservation in 1998 were helpful.

The assistance of chemist Dr David Spurgeon was invaluable. He offered insightful discussions, trouble-shooting support, drawings for the chemical equations, editing and general advice for the whole range of inquiry. Conservator Marilen Pool provided significant assistance in gathering relevant health and safety information and in reviewing draft tests. Conservator Thomas Braun retested numerous acid dilutions that resulted in a simplification of the concentrations recommended in the volume. Materials scientist Aniko Brezur offered metals of known assay for testing.

The award of a small grant from Dr Michael Cusanovich, Vice President for Research at the University of Arizona enabled the purchase of equipment and supplies, and funded the part-time assistance of conservator Matthew Crawford. Although the original goals of this project were modest, the encouragement of Dr David Kingery to expand the project and publish the research was of enormous value. The positive comments of conservation scientists Dr Eric Hansen and Dr Mary Striegel urged us to seek funds from the National Center for Preservation Technology and Training (NCPTT) under the direction of Dr Mark Gilberg. The participation of conservator Scott Carroll and chemist Dr David Spurgeon was partially supported with the NCPTT funds.

Mary Wood Lee, Director of the Campbell Center for Historic Preservation Studies, created a workshop and engaged our development of a curriculum for training of the spot-test techniques included in this volume. The opportunity to 'test the tests' was of tremendous benefit. We would like to thank the following colleagues from the courses who offered commentary that improved the tests: Jane Alison, James DeYoung, Suzanne Hargrove, Jane Ketcham, Margaret Little, Elissa O'Loughlin, Holly Lundberg, Ellen Pearlstein, France Remillard, Anne Witty, Dr Carol Aiken, Helen Alten, Dr Judy Bischoff, Sara Caspi, Helen Coxon, Molly Lambert, Christine Thede, Patricia Silence. In particular, we extend our very special thanks to France Remillard for introducing us to the use of the Pasteur pipette technique.

We are very grateful to conservator Chris Stavroudis and curator Dr R. Gwinn Vivian for their editorial suggestions. The enthusiasm and patience of our colleagues at the Arizona State Museum including archaeological collections curators, Mike Jacobs and Lisa Zimmerman; photographic archivists, Susan Luebbermann and Kathy Huebenschmidt; and photographer, Ken Matesich are also appreciated.

Finally, we would like to thank the selection committee of the Samuel H. Kress Foundation and the Foundation of the American Institute for Conservation for the Conservation Publication Fellowship. This award supported final preparations for this manuscript including reference editing by Linda Gregonis, text editing by Chris Stavroudis, and the illustrations by Stacey Lane.

1 Introduction

Historic and prehistoric objects of material culture found in museum collections or recovered by archaeological excavation are useful for interpreting past human behavior. Information can be extracted from the material of the artefact itself or from materials physically associated with the artefact, including accretions or deposits on the artefacts that qualify aspects of its surrounding environment. Much of this information becomes available to the conservator during artefact examination and treatment. However, some of this information may not survive treatment, so it is incumbent on the conservator to identify and record as much of this information as possible. The conservator is frequently confronted with decisions about conservation treatment that, dependent on material characterizations, requires access to expensive instrumentation or examination by a specialist. It is not feasible to carry out such treatment in most conservation labs. Less expensive 'spot tests' however, may provide the needed information, especially if they do not require elaborate instruments or specialist participation. Unfortunately, a directory of standardized spot tests for the wide range of artefacts and other materials that are of concern to the conservator, archaeologist, or museum specialist has not been available.

The characterization of distinctive qualities or characteristics of various materials begins with questions concerning the specific nature of an artefact material, technology, and aspects of its condition (such as deterioration). Specific artefact materials to be considered might include metals, ceramics, plasters, glasses, stone, pigments, bone, wood, fibers, gums, resins and accretions. Tests grouped for metals may identify hardness, color, alloying or anionic contaminants such as chlorides and sulfates. Tests for ceramics often correlate aspects of contaminant salts, temper, inclusions, or porosity to condition. Chemical tests are useful for distinguishing marble, plasters, shell, and calcareous accretions. Inorganic coatings or accretions may be identified by tests that differentiate carbonates, sulfates, and chlorides. Tests for organic coatings and adhesives may distinguish proteins, starches, gums, and resins. Specific tests for the materials that are contextual to the artefact material might include analysis of soil or ground water and dissolved solids. Various tests for soluble and semi-soluble salts and pH are useful in a number of contexts. The material characterization of conservation treatment supplies is also important and may include tests for purified water, adhesives, coatings, and storage materials (plastics, papers, writing materials).

Although various tests have been developed and recorded for chemical spot-test analysis, few characterize the composition of materials found in artefacts. The major developments and subsequent revisions of systematic chemical spot-test analysis by Fritz Fiegl (1937) and G. Vogel (1937) remain the important references for isolating organic and inorganic materials. Typically, these spot-test reactions have been successfully applied in clinical analysis, air quality testing, food and water analysis, geochemical prospecting, and in forensic laboratories. However, within the literature of these disciplines, the style of test formats varies considerably, thereby making comparative testing of the same materials by different tests difficult. Moreover, most of these tests are limited to chemical procedures and were designed for use by scientists who would not necessarily need a reminder of such aspects as reagent preparation, storage and safety, the chemical principle involved, how to interpret the test results, or the damage that can occur to the sample (artefact) as a consequence of the test.

Despite the rapid advances in, and the application of, modern instrumental analytical techniques such as chromatography, mass spectroscopy and electroanalytic analysis, the search for simpler, safer, and less expensive tests has continued to be important for many applications. In recent decades, much of the research on spot tests has been a review and elaboration of the old tests. The early work of Friedlieb Runge in 1842 (Anft 1953) involved the use of impregnated papers to accommodate a scientifically checked spot reaction and he is credited with producing the first spot-test papers. In the discipline of conservation, Rutherford Gettens and George Stout developed and refined numerous tests for the analysis of artist materials. Unfortunately, the result of most of their research on microchemical analytical techniques remains in the form of an unpublished notebook located in the Fogg Museum, Harvard University. Following the developments of instrumental analysis, a reliance on chemical characterization techniques declined. Today, because of the demand for easy, safe, and economical testing techniques, commercial chemical companies have been encouraged to produce simple and compact spot-test systems. Many of these special spot-test papers have applications in answering the questions posed by historic and artistic artefacts.

Although the theory of materials characterization has been discussed in recent years by conservators such as Beaubien (1994), the systematic assembly of methods for such characterization is commonly cited as a future direction for research. In general, tests designed specifically for artefacts are difficult to obtain because they are unpublished, or are not widely published in the conservation or archaeological literature. Most are either class notes or hand-outs, and tests for which the source is not the original reference, or in the form of partial photocopies of unreferenced publications. Successive generations of photocopied notations and texts frequently end up in the unwieldy format of a manila folder. Because much of the reference information is lost, forgotten, or, at best, difficult to access, many non-chemists are reluctant to take on the task of remembering a particular test and then dredging up the typically vague instructions to answer a relevant question. Also, because many test instructions for artefact applications are inaccurate, incomplete, or poorly written, there is a general mistrust of spot tests and a reluctance to rely on or even use them. The result is not only a narrower understanding of the materials, but a reliance on sophisticated instrumental techniques of analysis or outside personnel to answer many fundamental questions that could be answered with well defined spot tests.

Problems encountered with spot-testing guidelines have been particularly troublesome for museums. This results from a number of factors including a lack of good compositional descriptions for art objects or artefacts, a lack of written fabrication or technological records, and the fact that preservation planning decisions often need to be made on immediate material identifications. Therefore, this book is a response to problems faced in museums. A number of spot-test techniques (published and unpublished), used primarily by conservation professionals, were gathered and refined over a period of several years. The preliminary work involved a collaboration between conservation and chemistry. The process involved tracking down literature sources, selection of suitable tests (first round), and the development of concise protocols. Based on this process, a preliminary format for the systematic test documentation was devised. In 1991, a group of six tests that had been edited into this format were distributed in a workshop. The enthusiastic response of the workshop participants prompted interest in the development of a directory of spot tests for the professional conservator.

With a small grant in 1994 from the University of Arizona Office of the Vice President of Research, an expanded research project was initiated. The competitive award provided one-time start-up support for collaborative projects and enabled the hiring of a student assistant and the purchase of laboratory supplies. Funding from the National Center for Preservation Technology and Training (NCPTT) in 1995 supported:

(a) an assistant conservator to help evaluate, modify, and elaborate the tests;
(b) a consulting chemist to help clarify and interpret the chemistry; and
(c) the purchase of laboratory items such as an analytical balance.

Once the tests had been through preliminary protocol write-up, they were ready for actual experimentation and trial. In the autumn of 1995, the critical phase of the trials began and a true collaboration between the conservators and the chemists was realized. Because the tests are for non-chemists, ongoing discussions regarding applicability and ways to make them safer and easier to perform were necessary. An understanding of exactly what was being tested was critical at this stage. For example, a test to identify black pigments on pottery (Nelson 1975) actually identifies the iron component of black pigments in order to distinguish them from carbon-based pigments also found on pottery. Knowing that the identification of iron was the key concept made it apparent that there were broader applications for this test beyond pottery.

Modifications developed during the testing phase were done on a consultation basis with the two chemists to ensure that the tests remained effective. More importantly, it was during this phase that the initial vagueness of many of the tests was worked out. For example, questions such as "How much acid should be added and at what strength?" and "What if a weaker acid is used?" were clarified. When a test did not perform as expected, it was through the collaboration with the chemists and the conservators that the problems were finally resolved.

Important modifications were also made to many of these tests to make them less damaging to objects of art and archaeology. For example, the use of filter paper dampened with de-ionized water facilitated the capture of tiny pigment particles fallen from an object or found on a mount or in a storage box. Thus, the tiniest sample could be tested with no further loss of pigment or material from the artefact. Another technique involved using tiny, pinhead-size pieces of filter paper pulled from the edge of a larger sheet. Rather than placing chemical drops directly on the artefacts, the reagents are soaked into the small piece of filter paper, which is then placed on the artefact, and the results are viewed under magnification. This technique is particularly useful as most artefacts tend to have much more irregular surfaces than the normal test blanks used in laboratory experiments. Also, reagents that have low surface tension may flow readily over the surface of the object and spread from an inconspicuous test area to other areas, possibly causing damage to the artefact. By using the tiny filter paper modification there is greater control of the reagent.

Alternative test methods reported in the literature were also tried and were endorsed after repeated use and success during the trial phase. One example involved the use of a 6V battery to electrolyze a minute amount of sample material (literally picograms) on to a piece of filter paper that had been soaked in reagent. Although developed several decades earlier, this idea was published for use with artefacts by Marilyn Laver (1978). This electrolytic sampling method provides an excellent way to take very small samples and minimize the damage to the artefact surface. Although the battery set-up may seem intimidating, it is actually quite easy to perform and is a very reliable way to test metals.

Different apparatus, tools, and supplies were investigated and are recommended for the material characterization testing of small samples. Access to an analytical balance with sensitivity to 0.01 grams is essential for preparing reagents in small quantities. The requirement of laboratory instruments was avoided for this book, although the use of a microcentrifuge for many of the organic tests and a simple device for measuring melting point were included because of their more modest costs and their potential for additional testing applications. While observation of the reaction in some of the tests could be enhanced by stereo-zoom microscopic magnification, most of these tests do not require more than a 10× hand lens. Microchemical spot tests that require the magnification power of a microscope are another specialty and are beyond the scope of this volume.

The preferred supplies for comparing the reactions of a known positive, the unknown sample, and a known negative are spot-test plates. They are available in ceramic (white and blue or black) and in clear glass with varying numbers of testing wells. For other types of small samples, a jeweler's touchstone or the frosted end of a microscope slide can be useful. The innovative use of a sealed Pasteur pipette (Rémillard 1995) to contain and observe samples is an improved method for tests involving pyrolysis. Microtesttubes or capillary tubes may be used successfully to isolate the smallest of material characterization reactions.

Every effort has been made to elaborate on the interpretation of the test results. Generally, previous spot-test instructions fall short because they fail to identify or explain interference materials, or to elaborate on the various chemical processes and stages of a reaction. In addition, most of the published tests that report a color designation for a positive, neglect to indicate how long reactions may take or what negative reactions might appear to be. The tests in this book include those for metals, inorganic materials, organic materials, and general characteristics such as pH, specific gravity, and hardness. During the evaluation, writing-up, and trial phases, some of the tests were discarded because they repeated other tests, had limited application for objects of art and archaeology, or were deemed too difficult or dangerous to be performed in a typical conservation lab. We hope the compilation and translation of these tests for practical application to artefact characterization and conservation treatment decisions will be of value to several artefact-related disciplines.

2 Chemical safety

The decision to proceed with material characterization or chemical spot tests that involve the use of hazardous chemicals is a commitment to handle them responsibly throughout the processes of receipt, use, and disposal. The risks associated with handling and using chemicals are dependent on many factors, and it is not possible to define every possible health concern that may result from use or misuse.

A hazardous chemical may be defined as any chemical that poses a physical hazard or a health hazard. Many nations have specific definitions, regulations and standards regarding hazardous chemicals and these should be observed. As clarified by the United States Occupational Safety and Health Administration (OSHA), physical hazards include chemicals for which there is evidence that the chemical is a combustible liquid, a compressed gas, explosive, flammable, organic peroxide, oxidizer, pyrophoric, unstable (reactive), or water-reactive. Chemicals are health hazards if there is evidence that an acute (immediate) or chronic (delayed) health effect may occur following overexposure. Because overexposure depends on dose, duration, frequency, and the route of exposure, the health effects may be transient, persistent, or cumulative.

The preferred way to work with chemicals is to understand them, substitute less toxic chemicals whenever possible, implement safe work methods, install permanent engineering controls, and use personal protective equipment. It is considered unsafe to work alone in a chemical laboratory.

2.1 Understanding chemical hazards

Understanding chemicals is to use or handle, store, and dispose of them properly. The label is the most specialized format for providing important information regarding the physical and chemical properties of a particular chemical. Most labels contain an identification of the contents; the name and address of the manufacturer, distributor or importer and the appropriate hazard warnings. All hazardous chemicals should be stored in their original labeled containers and dated when received.

Chemical product labels

The example of a particularly informative sample chemical label reflects the range of important information. However, most labels are necessarily small in size and it is almost always necessary to obtain further information. The Chemical Abstract Service (CAS) is an excellent source and listing of the Chemical Abstract Service (CAS) number on the label refers to the systematic computerized chemical information source developed by the CAS Division of the American Chemical Society. Their publication, *Chemical Abstract*s is the largest repository of significant chemical research information reported in the international literature. The use of label pictograms, indicating if the chemical is explosive, oxidizing, flammable, toxic, harmful or irritant, corrosive, biohazard or dangerous for the environment, may be particularly helpful.

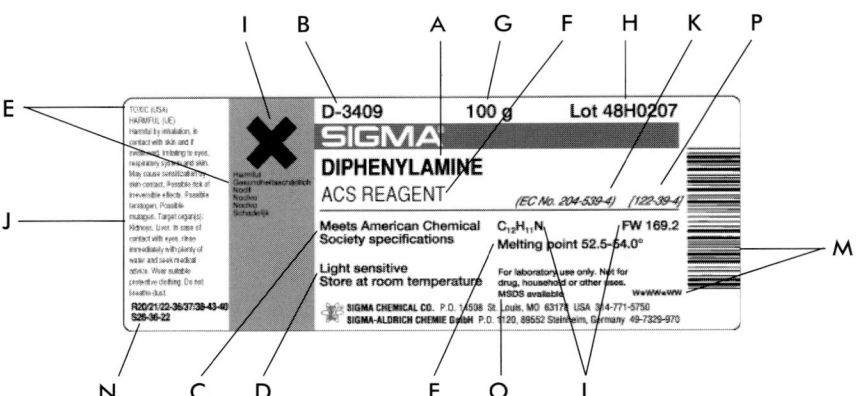

Key to Sigma product labels

    A Product name and description
    B Product number
    C Further descriptive information
    D Recommendations on handling and storage
        Storage temperatures indicated are for long-term storage of products. Products may be shipped under different conditions to reduce shipping costs, while still ensuring product quality.
    E Hazard statement
        Indication of danger.

F Lot analysis
  Data on activity, purity, degree of hydration, etc., for this lot.
G Package size
  Unless the material is described as pre-weighed, the package will normally contain at least the indicated quantity, and usually somewhat more. For some products, the actual quantity at the time of packaging is also shown. The user should always measure the amount needed from the container.
H Lot number
I Hazard pictogram
  Lets you know at a glance what safety hazards are involved in the use of this product.
J Further hazard information
  More complete description of actual hazards, handling precautions, and emergency management procedures.
K CAS number
  Chemical Abstract Service number shown wherever available. CAS numbers vary in how specifically they define the material. We make every effort to provide the most specific CAS number which applies. Where a CAS number is provided for a mixture or solution, it is usually the CAS number of the solute or component referred to in the main label name.
L Chemical formula and formula weight
  Unless water of hydration is indicated in the formula, the formula weight is for the anhydrous material.
M Bar code and eye readable equivalent
  The bar code and the eye readable equivalent of the bar code are for Sigma internal use and label identification.
N Risk and safety numbers
O Material safety data sheet available
  A material safety data sheet is available for this product.
P EC number
  This product has been identified with an EC number (EINECS or ELINCS). Those products without an EINECS number will carry the warning statement, "Caution – Substance Not Yet Fully Tested."

Pictograms

Pictograms are based on widely accepted standards.

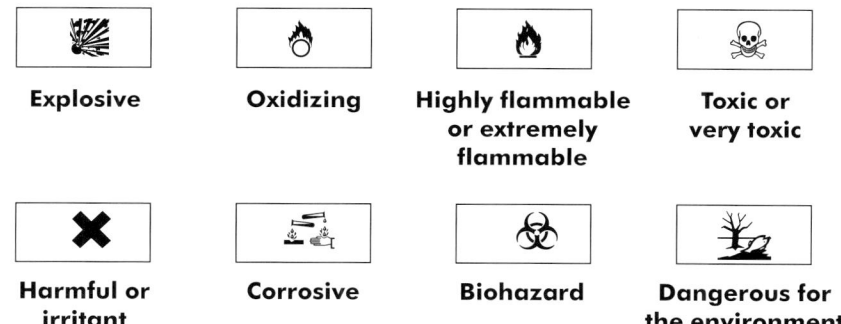

Materials safety data sheets

Most chemical supply companies will provide a materials safety data sheet (MSDS) or hazard statement for all the chemicals they distribute. The MSDS on any particular chemical is usually more informative than the label. Maintaining a notebook of MSDS sheets for all chemicals stored, used, or handled for spot tests is recommended. While the format for various MSDS sheets may vary from company to company, and some may not contain all of the same sections or categories, most will include similar information. A typical MSDS format must provide these sections.

Section One – Product information

Manufacturer's name, emergency telephone number, chemical name and synonyms, trade name, chemical family, formula.

Section Two – Hazardous ingredient information

Hazardous ingredients, percentages by weight or volume of each ingredient, Permissible Exposure Limit (PEL) and threshold limit values (TLV). Lower PEL or TLV levels are more hazardous.

Section Three – Physical data

Boiling point, vapor pressure (higher vapor pressure is more hazardous), vapor density, solubility in water, appearance and odor (a general order of low to high hazard chemical forms is solid>granules>dust, liquids>mists, smoke/fumes>vapors/gases), pH (chemicals of intermediate pH are less hazardous than those with low or high pH), specific gravity, percent volume by weight, evaporation rate.

Section Four – Fire and explosion hazard data

Flashpoint and method used (flashpoints vary according to how they are obtained and the method used is important, chemicals with a low flashpoint are a greater flammablity hazard), flammable limits, extinguishing media, special fire-fighting procedures, unusual fire and explosion hazards.

Section Five – Health hazard data

TLV, effects of overexposure (toxicity data is indicated in the form of average lethal dose LD50 or lethal concentration LD50 and low levels are most hazardous), emergency and first-aid procedures.

Section Six – Reactivity data

Stability, conditions to avoid (such as heat, shock, cold, light), incompatibility, hazardous decomposition products, hazardous polymerization.

Section Seven – Spill or leak procedures

Methods for control and clean-up, waste disposal method.

Section Eight – Special protection information

Respiratory protection, ventilation, protective gloves, eye protection, other protective equipment (such as apron, boots, glove box).

Section Nine – Special precautions

Additional precautions to be taken in handling and storage.

Reference guides such as *Prudent Practices in the Laboratory* (National Research Council 1995) or *Dangerous Properties of Industrial Materials* (Sax and Lewis 1996) provide valuable compilations of information on safety hazards. They generally emphasize the laboratory use of chemicals rather than the transport or handling concerns and more commonly are related to the industrial manufacture, distribution and use of these. Taken together, the chemical product labels, Material Safety Data Sheets, and various chemical guides and dictionaries (e.g. The *Merck Index*) provide knowledge of a chemical's physical, chemical, and biological properties; information on its proper storage and shelf life and guidance on the preferred manner of disposal.

2.2 Controlling chemical hazards

Controlling the hazards may require:

- engineering modifications for isolation, minimization, and ventilation
- administrative controls for planning, policy and procedure development, training and medical surveillance
- personal protective equipment (PPE) for individual protection.

2.2.1 Engineering modifications

Laboratory working conditions should include ventilation that provides workers with comfortable breathing air. Non-laboratory room ventilation levels are not usually sufficient for preventing accumulation of chemical vapors. It is recommended that all work carried out with chemicals with low TLVs or high vapor pressures be done in a fume hood (fume cupboard).

The primary purpose of the laboratory fume hood is to prevent undesirable vapors from entering the general laboratory atmosphere. The successful performance of a hood depends primarily on the velocity of air moving through it. This is affected by cross-drafts, entrance shapes, thermal loading, mechanical action of particles and obstructions of the hood. Hood users should follow procedures that do not hinder the function of the hood.

With the sash down, the hood places a protective barrier between the worker and the chemical operation. The hood will minimize the effects of explosions, fires, spills or similar events that may occur within the hood.

The sash position that indicates the velocity for adequate containment (not allowing vapors to spill out, powders to transfer, or flame control to be affected) should be labeled on each hood. Sash height should be high enough to allow normal work to proceed.

If a hood is available:

DO
- prepare all toxic chemical reagents inside the hood with labels
- operate the hood at the proper sash height (which should be labeled and dated)
- keep the sash or shield lowered at all times except when adjusting the apparatus inside
- keep the hood on at all times when a chemical is inside the hood, regardless of whether any work is being done in the hood
- work at least 10cm behind the front edge of the hood
- have only the equipment necessary to conduct the experiment in the hood
- keep vent ducts and fans clean and clear of obstructions
- use equipment or apparatus that have legs to raise them off the work surface and allow even airflow
- place equipment and apparatus toward the rear of the hood to prevent vapors from escaping
- reduce clutter and temporary storage to a minimum
- minimize pedestrian traffic in front of hoods, particularly during hazardous experiments
- know the toxic properties of the substances you are working with and be familiar with the signs and symptoms of overexposure.

DO NOT
- store chemicals or equipment in a hood where work is conducted
- stick your head into the hood during experiments
- position equipment or other materials so as to block rear hood slots or otherwise interfere with airflow
- position fans or air conditioners in a manner that will direct airflow across the face of the hood and interfere with containment.

Fume scrubbers or absorbers are portable self-contained hoods designed for absorbing odors and small quantities of solvent and acid fumes through an activated carbon filter. The blower draws fumes and vapors through the filter where they are trapped before clean air is returned to the room. These types of units should be checked annually (isoamyl acetate or banana oil may be used to determine if the filter is functioning properly or has been used up). Also, contaminated filters require disposal procedures for hazardous wastes.

Additional safety equipment that is important for chemical laboratories include: emergency eyewash stations, safety showers, first-aid stations and fire extinguishers. These devices require regular inspection and should receive a tag or label indicating when the last inspection occurred.

2.2.2 Administrative controls for chemical hygiene

All laboratories planning on conducting chemical spot tests should have a chemical hygiene plan. The chemical hygiene plan is the primary document for administrative planning, policy and procedure development, training and medical surveillance. Its primary purpose is to ensure that laboratory employees and volunteers are educated about the potential hazards of working in a chemical laboratory before they undertake any tasks that could expose them to the hazardous chemicals. Chemical hygiene plans were formally introduced in the United States in 1983 when the Occupational Safety and Health Administration (OSHA) of the United States published the Hazard Communication Standard, known as 29 CFR 1910.1200, which applied to certain manufacturers and in part to certain laboratories. The 'HazCom' Standard required employers to acquire and make available to their employees Material Safety Data Sheets. Despite the unique differences in various laboratories, including the amounts of chemicals used and the variety of chemicals used, OSHA proposed a rule to require chemical hygiene plans.

On 31 January 1990, the United States Department of Labor published an amendment to 29 CFR 1910, Subpart Z, identified as the Occupational exposure hazardous chemicals in the laboratory or Section 1910.1450. The title of that amendment is better known as the 'Laboratory Standard', and the development of a chemical hygiene plan for chemical laboratories in the United States is a requirement of that standard.

According to OSHA, the chemical hygiene plan requirements apply to the laboratory use of chemicals on a laboratory scale or non-production level. They refer to laboratories as facilities where relatively small quantities of hazardous chemicals are used. Laboratory scale refers to the work that is done with chemical substances in which the containers used for the reactions, transfers, and other handling are designed to be easily and safely manipulated by one person. Based on these guidelines, most conservation laboratories are also chemical laboratories.

A laboratory chemical hygiene plan is designed as a tool for coordinating safety procedures. The benefits of a plan are the prevention of injuries and illness through an increased worker awareness of potential risks, improved work practices, and appropriate use of existing personal protective equipment. It is assumed that the more informed workers are about potential hazards and emergencies, the more they will take the necessary precautions to prevent them. Standard chemical hygiene plans usually outline several categories of information depending on their size, structure and program. A typical format for use by a museum conservation laboratory might include:

Section 1 – Introduction

Purpose of plan, history of chemical safety.

Section 2 – Description of laboratory

Description of laboratory program or activities, description of relationship to larger organization, floor plan of structure.

Section 3 – Responsibilities for the chemical hygiene plan

Personnel responsible for plan, joint responsibilities, annual reviews, responsibilities of administrators, supervisors, employees, volunteers, or other departments.

Section 4 – Information and training

General and specific sources and opportunities for training and information (i.e. first-aid, cardio-pulmonary resuscitation (CPR), fire extinguishers).

Section 5 – Standard operating procedures

General safety rules, procedure-specific safety procedures (i.e. hazardous waste disposal), chemical control measures (equipment operation and maintenance), emergency stations and personal protection apparel.

Section 6 – Criteria for control measures

Exposure guidelines, fire guidelines, reactivity guidelines, and corrosivity and contact hazards.

Section 7 – Exposure evaluations and medical surveillance

Validation of safety equipment, laboratory chemical hygiene self-evaluation, formal laboratory inspection, exposure evaluations or monitoring, medical consultation, documentation.

Section 8 – Emergency procedures

General information, emergency preparedness plan, medical care, interim first-aid, fire, chemical spill, shutdown of dangerous activity.

2.3 Personal protective equipment

The success of personal protective equipment (PPE) in day-to-day work or in a chemical emergency depends on its proper fit, use and care, as well as its inherent chemical limitations and life expectancy. Use of the following items is recommended while conducting chemical spot tests.

Goggles/safety glasses

The eyes are perhaps the most fragile parts of the body. Eye protection is recommended during chemical characterization tests when splashed chemicals, flying glass, sparks, noxious fumes, vapors and gases or light radiation can cause permanent damage to the eyes. Impact- and splash-resistant goggles that are the equivalent or better than the American National Standards Institute approved (ANSI Z87.1-2003) safety glasses are basic equipment and should be worn at all times when there is potential for exposure to laboratory hazards. (Other countries may have their own standards and these should be carefully checked.) Goggles should be worn working with liquid chemicals as a liquid-proof seal around the eyes is necessary. Safety glasses or spectacles should only be used when working with solid materials and not with liquid chemicals or solvents. Ultraviolet inhibitor spectacles or goggles made of lightweight plastic should be worn if using the ultraviolet examination lamp, particularly if short-wave radiation is being used.

Gloves

The hands are the part of the body most likely to come into contact with chemicals during chemical spot testing. Some chemicals go through the skin and can cause illness in other parts of the body. Wearing gloves helps to protect workers from skin irritation and other effects of chemical exposures. Maximum hand protection is the result of a careful matching of applications and working conditions with glove composition and characteristics. Use of damaged, worn or the wrong type of glove can actually increase the hazards because skin absorption is enhanced through the confined skin contact within the contaminated glove.

The glove should be selected to match the potential hazard. Nitrile, neoprene, vinyl plastic, synthetic-latex, and rubber-latex materials vary widely in chemical resistance (see chart below). It is important to evaluate the physical properties of the material. In addition to chemical resistance, materials vary in physical toughness – what may be safe with one chemical may prove to be harmful with another. Gloves used for especially dangerous chemicals can be tested for leaks by filling them with air and immersing them in water. Gloves used with highly toxic materials should be disposed of as hazardous waste before leaving the work area.

Glove chart

Chemical	Nitrile	Neoprene	Vinyl plastic	Syn-latex	Natural latex	Rubber-latex
Alcohols	Excellent	Good	Excellent	Excellent	Excellent	Good
Caustics	Excellent	Excellent	Excellent	Excellent	Excellent	Excellent
Chlorinated solvents	Excellent	Good	Fair	Good	Not recommended	Not recommended
Ketones	Good	Good	Not recommended	Good	Good	Good
Petroleum solvents	Superior	Excellent	Good	Excellent	Fair	Fair
Organic acids	Excellent	Excellent	Excellent	Excellent	Excellent	Kxcellent
Inorganic acids	Excellent	Excellent	Excellent	Excellent	Excellent	Excellent
Non-chlorinated solvents	Good	Good	Fair	Good	Not recommended	Not recommended

This table is based on information offered by various safety supply companies.

Lab coats and aprons

Aprons, lab coats, jump suits and oversleeves are among the kinds of apparel that may be useful or even necessary in a chemical laboratory. Thought should be given to clothing as many fabrics such as nylon and polyester react with common lab chemicals and burn easily. Laboratory coats that are long-sleeved offer the wearer some skin protection against minor splashes by allowing the chemical to react with fabric before reaching the skin, thereby offering the victim time to remove the coat and shower. Aprons also offer additional time to shield and protect the wearer from chemical splash. The use of sleeve protectors or arm guards affords added arm and garment protection above the gloves. The use of closed-toe shoes is recommended because broken glass and chemical spills are possible.

Respirators

A respiratory-protective device is used to protect the wearer from the inhalation of harmful atmospheres. It should be kept in mind that the use of respirators is sanctioned only when engineering and work-practice controls are not feasible or are inadequate. Respiratory-protective devices are grouped broadly according to their mode of functioning. These classifications are:

- air-purifying respirators which remove the toxic contaminant from the air by a filter
- atmosphere-supplying respirators which supply clean air from an airline or tank.

Selection of the proper respirator requires the careful consideration of the following factors:

- The chemical, physical and toxicological properties of the substance against which protection is required. This includes hygiene standards, flammability levels, warning properties, eye and skin irritation and the possibility of oxygen deficiency.
- An evaluation of actual and potential exposure concentrations.
- The nature of the duties to be performed by the wearer of the respirator, particularly as it relates to restriction of movement and period of time of use.
- An understanding of the principles, design, scope of use, limitations, advantages and disadvantages of the respiratory-protective equipment available. Respirators that use cartridges or canisters can be rendered entirely worthless if the correct filter is not used.

Before using a respirator, the user should be fully aware of the various types available and the conditions under which they should be used. Both OSHA and ANSI require that respirators be tested for effectiveness of facial seal. Training in the use of respirators should include:

- instruction in the nature of the hazard and what may happen if the respirator is not used
- explanation of why more positive control is not immediately feasible
- a discussion of why this is the proper type of respirator, its capabilities and limitations
- provisions for maintenance which should include inspections for defects and leaks, cleaning and disinfecting, repair and storage.

Non-toxic particle masks are disposable masks used for exposures to nuisance levels of dusts and powders below the current maximum level of

15 mg m^{-3} as defined by OSHA. They are not to be used for mists containing gases, vapors or non-absorbed contaminants.

2.4 Conclusion

By nature, chemical laboratories can be dangerous places to work. An exposure to hazardous chemicals presents a significant health risk to those conducting chemical-characterization tests – use of the wrong chemicals or misuse of laboratory techniques could result in a serious accident with severe or permanent bodily damage. Laboratory safety should also take into account that reagents (mixtures of chemicals) may have working properties and hazard concerns that vary from the chemicals that compose them. It is imperative that all tests be performed as safely as possible.

3 Scientific method and techniques of spot testing

The purpose in utilizing material characterization-testing techniques is to establish the composition, identity or nature of naturally occurring or artificially manufactured substances. Within the discipline of analytical chemistry this is usually done with two distinct approaches: qualitative analysis utilizes methods that chemically identify materials; quantitative analysis utilizes methods that determine the percentage of composition or constituents in materials. The material characterization tests included in this volume represent non-chemical techniques as well as inorganic and organic methods of qualitative chemical analysis. This book does not include the newer, faster, and more specifically accurate techniques that are collectively called instrumental analysis.

Feigl (1972: 31) refers to spot-test analysis as a generic term applied to sensitive and selective tests based on chemical reactions whereby the use of a drop of the test or reagent solution is an essential step. Spot tests can provide relatively quick and easy answers to material characterization questions. Sometimes they satisfy the immediate need for information, at other times they direct the application of further testing by methods of instrumental analysis. Chemical spot testing is rarely carried out on samples of completely unknown character – usually some previous knowledge or information about the sample and its context is known.

The best approach to conducting material characterization tests for objects of art and archaeology is to follow the guidelines of experimental science and to creatively utilize and adapt the scientific apparatus and supplies of the laboratory. The tests in this book are not organized as flow charts of successive steps, however, the design of test frameworks or programs of action that utilize multiple tests is possible by combining and modifying the tests.

3.1 Scientific method

Scientific method begins with a problem or question that is raised by observation. For objects of art and archaeology this may include questions regarding composition, deterioration products, accretions or deposits, or the nature of associated materials such as packaging. Spot-test techniques may not provide definitive or absolute answers to these questions but they can narrow the possibilities and perhaps eliminate certain classes of materials altogether.

The hypothesis or prediction guides the design of an experiment – aimless experiments are almost always unproductive. A certain level of practical and theoretical knowledge of chemistry is required to intelligently utilize spot-testing techniques. To apply spot-test techniques to objects of art and archaeology, background in the technology of material culture is also required in order to recognize and distinguish the problems that may be addressed. A search of the relevant literature is necessary if familiarity in these areas is lacking.

Experiment is the basis of scientific method and bias is avoided by utilizing good experimental design and by not regarding the tests as recipes. Useful observations may be made by carefully controlling the conditions of each spot-test formulae. These controls include the use of:

- experimental trials
- comparative observations
- accurate record keeping
- good laboratory hygiene
- efficient laboratory technique.

One of the benefits of conducting controlled experiments is repeatability which enables greater confidence in both the techniques and their results. Another phenomenon of conducting spot-testing experiments is that sometimes entirely new problems are raised during an investigation. Successful researchers take advantage of all results, record them, and draw conclusions from their experiments.

3.1.1 Experimental trials

Simple spot-test techniques of material characterization involve dropping a reagent solution onto an unknown substance and observing the reaction. The significant part of conducting spots tests under controlled conditions relates to the actual manipulation of the reagents with the unknown substances so that maximum sensitivity and selectivity may be obtained. It is important to carry out the test several times in order to have confidence in the results.

3.1.2 Comparative observations

Spot-testing techniques rely on comparisons. It is important to utilize blanks of known negatives and known positives in observation trials. A blank contains no sample material and is used to determine whether the chemical reagents give any kind of reaction by themselves and to establish that they have not been contaminated or decomposed. It is also useful to compare the appearance of the blank to see how much of a difference there is to the unknown sample. Known negatives are samples of material that are of known composition and that are expected to produce negative results. They ensure that the test result is not a false positive and also provide a comparative color value, especially in samples that are similar to each other. Known positives are samples of material of known composition and which are expected to produce positive results. They ensure that the chemicals and the technique are working properly and permit observation of a positive result. It is recommended that the knowns used for comparative testing be materials that are similar in composition or that might be confused with the material being tested. For example, if aluminum is being tested, then other white metals would serve as good known negatives.

3.1.3 Accurate record keeping

Spot testing requires questioning, checking and verifying the data obtained by experimentation so record keeping is essential. Experimental details obtained during spot-test trials should not be taken for granted, rather they should be recorded clearly, accurately and objectively. The data can be recorded in a simple notebook or into a computer database, but records should be made during the experiment regardless of the format. An important aspect of any recording format is the date – this single item helps interpret the experimental results in light of further knowledge.

A computer database is more convenient for recording large numbers of tests or for retrieving prior test information quickly. Fundamental information to record includes:

- information about the material being tested (source, weight, color and how the sample was obtained)
- the name of the test used to characterize the material
- reagent information (reagent weights and volumes, dates of preparation, temperatures or general environmental conditions, and techniques that describe the way the test was performed)
- observations (a record of what reactions are seen)
- remarks (interpretation of the results).

An accurate record of the test trials provides valuable information for future reference. If all of the information, including the results of blanks, known negatives and known positives, is recorded it becomes a refresher guide when the exact procedures and results have been forgotten. Review of past test records can help to trouble-shoot or diagnose a problem. It is recommended that all details, even if they seem minute or trivial at the time, be recorded (see example of sample trial record form in Appendix 7).

3.1.4 Good laboratory hygiene

Good laboratory hygiene is the key to obtaining reliable and repeatable spot-test results. The space devoted to spot testing must be kept as clean and tidy as possible. Glass and porcelain-ware should be cleaned immediately after use because cleaning is easier with fresh residues which have not hardened or dried. The use of a cleaning product designed for laboratory glassware that does not leave a residue should be followed with a rinse of de-ionized or distilled water. Some spot tests are so sensitive that tap water that has dried on the surface will cause false positive results. Never use cracked or broken glassware – it may cause the sample to be lost, cause injury and contamination. Be careful when storing glass and porcelain-ware to avoid breakage when drawers are opened and closed.

Many spot-test reactions require the use of acid solutions. Acids are corrosive: they will burn, irritate or destructively attack the skin. Particular suggestions for using them include:

- Always wear gloves, goggles and protective clothing when working with acids or bases.
- **Always add acid to water.** This is important because the dilution action is exothermic (heat is evolved) and spattering may occur. When pouring, grip the label area on a bottle to avoid touching any drip residues that may remain from previous pouring.
- Keep the number of stock dilutions and their quantities to the smallest practical minimum.
- Store the dilutions in labeled, small acid-resistant containers out of direct sunlight and at or below benchtop level.
- Follow local chemical waste regulations and manufacturer's MSDS guidelines when disposing of all chemicals. Very small quantities of common acid or base dilutions may usually be poured down the drain if they are neutralized and/or added to a significant amount of water (see Figs 1 & 2).

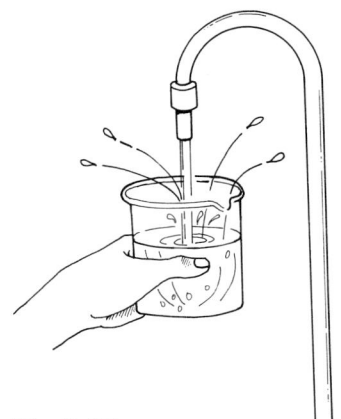

Fig. 1 Wrong

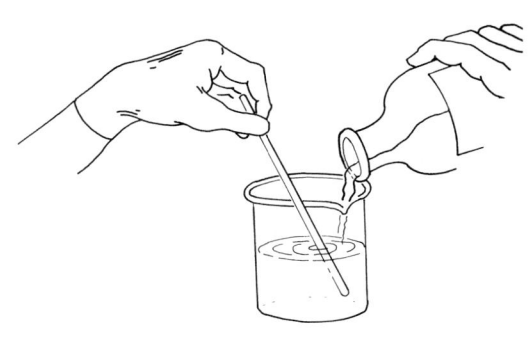

Fig. 2 Always add acid to water

3.1.5 Efficient laboratory technique

Laboratory needs for spot testing depend on the types of tests performed. The actual amount of space required can be quite small as the most appropriate scale for conducting material characterization tests on objects of art and archaeology is semi-micro. By using small quantities of sample material, the amount of chemicals, bench space and the time necessary for the experiments are also reduced. A supply of reagents does not have to be large, but the variety is extensive. Chemicals of analytical-grade quality are recommended (the grade refers to a purity standard). While other grades of chemicals may work, the presence of trace impurities is a problem in some tests. All chemical reagents and test solutions should be labeled clearly. It is also desirable to have representatives of diverse classes of artefact materials for use as known positives and negatives in the trials. A collection of accurately characterized plant and animal products, synthetics, minerals, metals and various technical products can be of great value in the preliminary examination and testing of unknown materials.

Spot-testing techniques invite variation and refinement. For example, a test for the presence of chromium may be used to distinguish chrome metal, chromate pigments or chrome-tanned leather. It is important to understand what exactly a particular test is detecting, and to use techniques that offer the best match of test question to test material. Most of the chemical methods used in the characterization of artefacts employ one of the following processes:

- A drop of reagent is placed directly but unobtrusively on part of an object or on a small sample removed from the object.
- A drop of reagent is placed on a piece of filter paper that has been previously impregnated with sample test solution from an object.
- A drop of reagent is placed near a drop of test solution or test sample on a supporting surface, such as paper, glass slide, porcelain, or stone and the two are drawn together.
- A strip of filter paper or a drop of reagent is subjected to the action of electrolysis involving the test sample (metal).
- A drop of reagent is added to a small volume of test solution and the reaction products are extracted with organic solvents.

A variety of glass, porcelain and plastic-ware items are used in conducting chemical spot tests. Assorted sizes of glass beakers, graduated cylinders, Erlenmeyer flasks, watch glasses, funnels, pipettes, test tubes, acid-resistant storage bottles, screw-capped vials, capillary tubes and spot plates are necessary. Glass-stirring rods, microscope slides and Pasteur pipettes are also important. Porcelain spot plates in white and black or blue are the appropriate reaction seat for many of the tests. Polyethylene pipettes, weighing pans and papers, and storage bottles are particularly handy because they can be discarded after use. Droppers, Pasteur pipettes and pipettes are not the same. The dropper is a non-quantitative means of transferring a liquid from one container to another and 20 drops is about equal to one milliliter (mL). Pasteur pipettes are better for manipulating smaller volumes of liquid and 40 drops is approximately equal to one mL. Pipettes are calibrated and are used when exact volumes must be transferred. Tweezers, spatulas and clamps of stainless steel, nickel-plated steel or Teflon-coated steel are fundamental for manipulating the sample material and various supports during testing. However, they can contaminate or negate certain tests, particularly the metal and electrolysis tests (see Fig. 3).

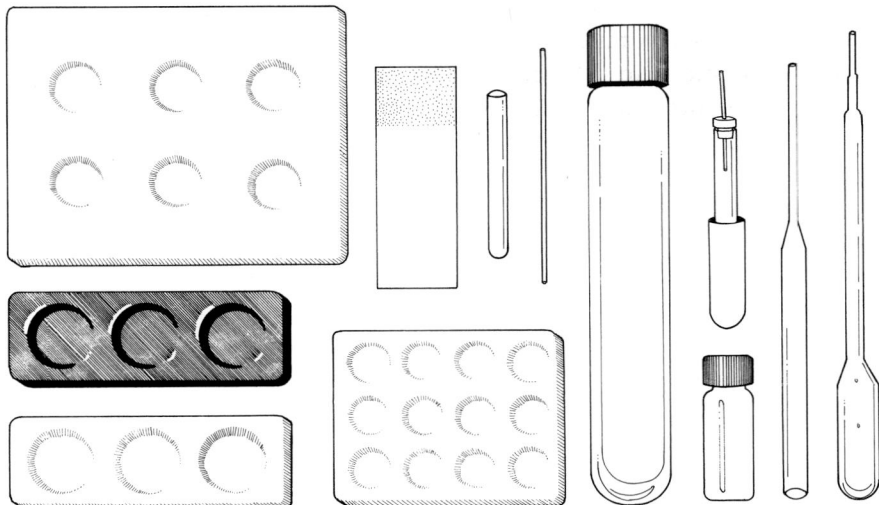

Fig. 3 Supplies for spot testing

Test papers are an important component in chemical spot testing. Those most useful include filter papers, pH-indicating papers and commercially produced reagent-impregnated dry papers. Strips, wedge-shaped pieces and bits of qualitative filter paper (Whatman No. 3) are the most common types used and can be stored in petri dishes or vessels with airtight lids. This paper leaves a small amount of ash upon burning and may contain traces of iron and phosphate. When testing for these reactions the use of quantitative filter paper (Whatman No. 42) or a hardened filter paper that is lintless (Whatman No. 542) is recommended. The use of impregnated-filter paper facilitates the use of chemical spot tests on objects of art and archaeology by limiting the amount of contact between the artefact material and the reagent. When used for testing, the paper is held to the object with tweezers. It is important to note that the tweezers should not touch the wet area (except for reactions involving electrolysis) because alloys in the tweezers can affect the test results. Sometimes plastic or wooden tweezers or a split bamboo skewer may be the best choice (see Fig. 4).

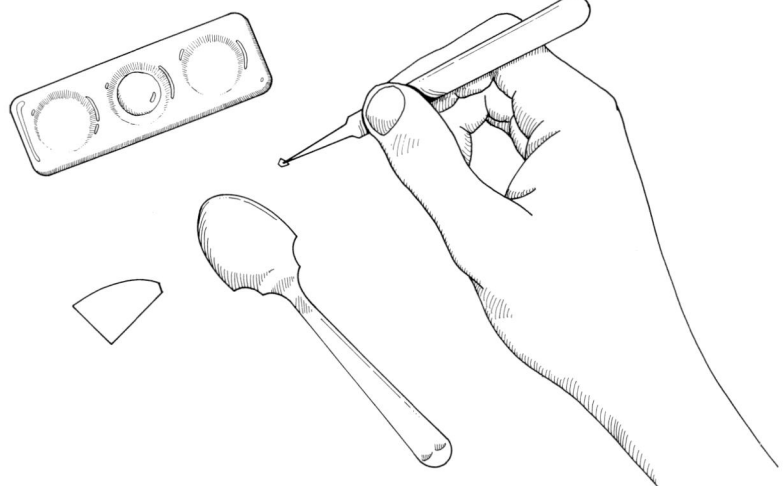

Fig. 4 Use of tiny bits of paper for chemical spot tests

Papers that are pH-indicating measure the hydrogen-ion concentration of a solution by incorporating an indicator in the paper and attaching it to a semi-rigid plastic strip. The color of the acid-base indicator in pH papers changes in the presence or absence of an acid or base concentration. The pH-indicating paper strips are particularly useful for characterizing the evolution of gas or vapor from a reacting sample. A useful technique involves folding the strip so that it springs open and remains positioned by resistance at the top of the test container, normally a test tube or Pasteur pipette (see Fig. 5).

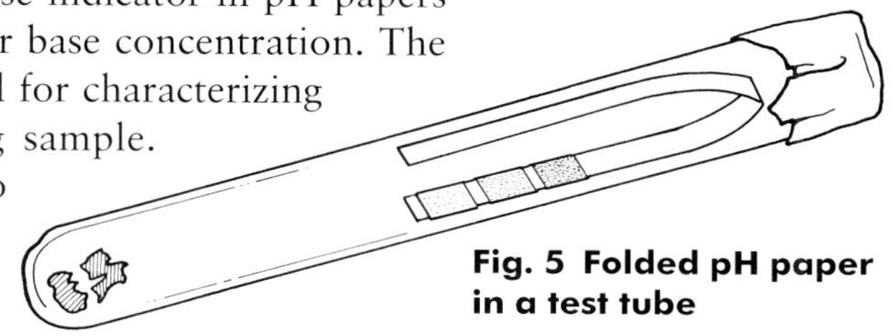

Fig. 5 Folded pH paper in a test tube

Many commercial companies have developed strip tests in which papers are impregnated with reagents and other substances. Each paper strip serves as the reaction zone for a specific spot test. Spot-test reagent papers offer a safe, easy and portable variation to chemical spot-test analysis and are excellent for field applications. Supplied as simple paper strips, disks or attached to semi-rigid plastic they can be conveniently transported and stored. They may be cut into smaller shapes and applied to very small samples or to objects of unusual form.

Sometimes a porous porcelain (unglazed ceramic), glass (frosted end of a microscope slide) or a jeweler's touchstone (a smooth, hard, black stone) are used to seat a chemical spot-test reaction involving a fine deposit of sample material and a drop of reagent solution. For example, the jeweler's touchstone is used in conjunction with a set of test needles to test gold or silver alloys. Each needle is tipped with a known-quality gold or silver alloy and is stamped with a quality number. To characterize an unknown gold or silver alloy, the needles are used to make parallel strike lines that deposit a small amount of alloy by minute abrasion onto the stone. It is easier to see the comparative reaction if the strike line from the unknown sample is made between the knowns. If a large drop of reagent acid is allowed to flow simultaneously over all the strike lines, the reaction may be observed across all lines. Interpretation of the results is comparative and the process may involve repeating the test with different strike combinations. It is important not to let the dropper, pipette, or glass rod contact the metal strike lines as this may contaminate the reagent. The metals are not absorbed by the stone, but it is critical that the stone be cleaned after use and that acid reagents not be allowed to dry on its surface as the residue may interfere with future test accuracy (see Fig. 6).

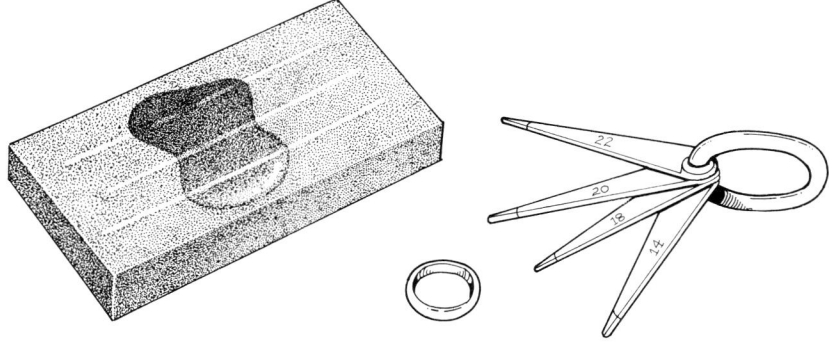

Fig. 6 Jeweler's touchstone and karat needles

3.2 Auxiliary laboratory equipment

Accurate auxiliary laboratory equipment is also important to chemical spot testing. For example, a balance is essential for most of the tests included in this book. In addition, many of the tests require a source of heat or flame that is used in combination with specialized characterization techniques. Some of the tests utilize contrivances such as the set-up for sampling by electrolysis or small machines such as a centrifuge or a melting point device. A magnet, magnification glass or stereo-zoom microscope are useful items to have on hand while performing many tests. In addition, photographic processing is necessary to characterize radioactivity in objects. All laboratory devices are delicate, so when applicable, the manufacturer's instructions for care and operation should be followed to ensure safe and accurate performance.

3.2.1 Balances

In addition to making volumetric measurements of liquids it is necessary to make mass measurements of solids. The mass of dry material is determined in the lab by weighing with a balance. The choice of a beam balance, a mechanical balance or an electronic analytical balance depends on the degree of sensitivity required. A modern electronic analytical balance is extremely sensitive and has a digital display of up to four decimal places. When spot tests require measurement of reagent mass or weight with an accuracy of 0.05 grams or less an electronic analytical balance should be used. A sensitivity of 0.1 implies that there are two places after the decimal – sensitivity in this case refers to the ability to weigh a material to that amount.

Most electronic analytical balances are very sensitive and have a glass housing covering the pan to keep air currents to a minimum. For use, the pan should be clean and the doors should be closed. All balances need to be placed on a surface that is level and not affected by outside vibrations, and the leveling and calibration should be checked regularly. Most modern instruments have a self-calibration procedure upon being turned on.

3.2.2 Specific gravity measurement

Measurements of specific gravity require the use of a balance. Specific gravity is a numerical value related to density; it is the weight of a substance divided by the weight of an equal volume of water. A simple device for measuring specific gravity may be purchased or constructed. A device made largely of foam, foam board or cardboard can consist of:

- a hanging support with a hook that rests on the balance pan
- a platform that rests above the balance pan
- a small foil basket with hanging wires that can be submerged into a beaker of water while on the balance pan.

Specific gravity is determined by taking multiple measurements (at the same temperature) of the device (in air and in water) and calculating the ratio (see Fig. 7).

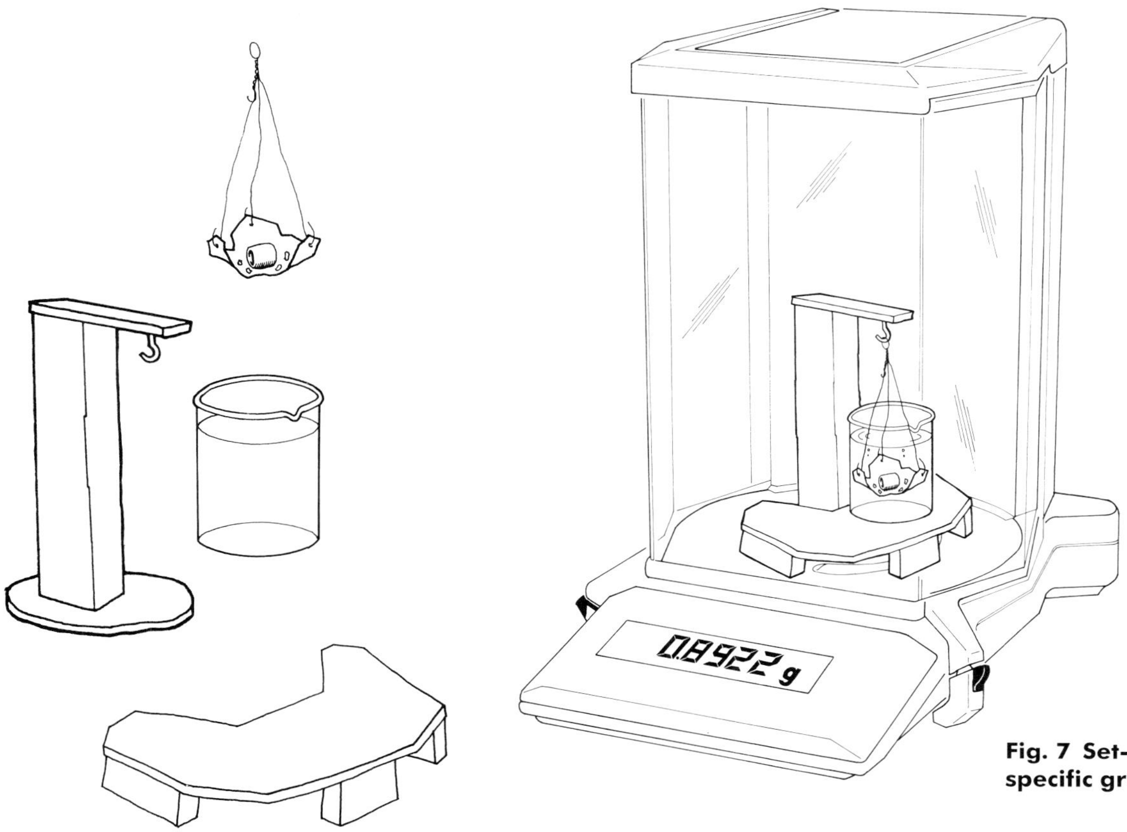

Fig. 7 Set-up for specific gravity

3.2.3 Sources of heat or flame

Some spot-test instructions call for sources of heat. Bunsen burners, alcohol lamps, safety (cigarette) lighters or hot plates may be used. Practically all organic liquids are flammable so it is important to always check the label, MSDS or a reference book for the boiling point prior to using a heat source. The flame of a Bunsen burner consists of three parts: the reducing flame or the inner blue cone of mostly unburnt gas that is about 320°C; a luminous tip (visible when the air holes are slightly closed); and an oxidizing flame or the outer mantle, where complete combustion of the gas occurs. The hottest portion of the flame has a temperature of about 1560°C and is located about one-third of the way up the flame, approximately equidistant from the inside and outside edges of the mantle.

If an alcohol lamp is used, only approved fuel such as denatured alcohol should be burned and lighted burners should never be carried. If using a butane or cigarette lighter (for the Beilstein test, p. 29) remember that the lighting mechanism can become extremely hot after prolonged use.

Heating liquids over a flame can be very dangerous and the following suggestions are advised:

- Never fill the test tube or other container more than one third full.
- Always hold the test tube with a clamp.
- Avoid making a flame any larger than absolutely necessary.

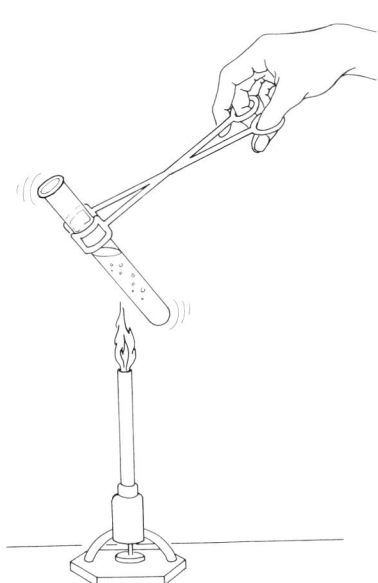

Fig. 8 Heating a test tube

- When placed in the flame keep the test tube moving constantly (if the test tube is not moved, hot vapors will form and eject the liquid violently – this is called bumping).
- Always point the test tube opening away from yourself and anyone else in the area (see Fig. 8).
- Always use extreme caution when applying heat to a closed container (for example screw caps should be loosened).

The choice of heating system may depend upon the flammability of the liquid, the type of vessel or container used, or the presence of fire hazard. In some tests, a hot-water bath using a beaker of water on a hot plate is recommended. For example, small samples in vials or centrifuge capsules cannot be heated over an open flame because of bumping or spattering of the liquid (see Fig. 9).

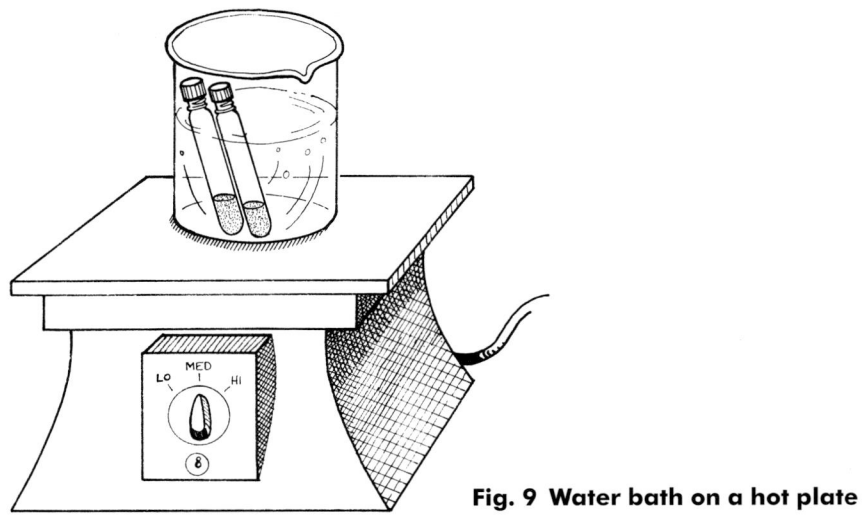

Fig. 9 Water bath on a hot plate

3.2.4 Testing techniques using pyrolysis

Pyrolysis – the use of heat alone to cause the breakdown or transformation of a compound – can be an applicable technique in many material characterization tests. During the process sublimation (the direct passage from the solid state to the vapor state), melting or decomposition (with heat evolved or absorbed) may occur. Two pyrolysis techniques can be adapted for use with objects of art and archaeology: the Beilstein test for chloride uses a fine copper wire and the Remillard technique uses Pasteur pipettes.

In the Beilstein test, materials containing chloride, such as poly(vinyl chloride), may be melted onto a copper wire that has been heated and then placed onto the test sample. When the wire, with chlorides from the sample, is heated, copper chloride forms and volatilizes, imparting a characteristic green color to the flame. The wire may be cleaned by burning and wiping the residue off (see Fig. 10).

The decomposition products of pyrolysis may produce colorless or colored gases, a characteristic smell or a specific pH. The Remillard Pasteur pipette technique affords an improvement over the use of test tubes by containing

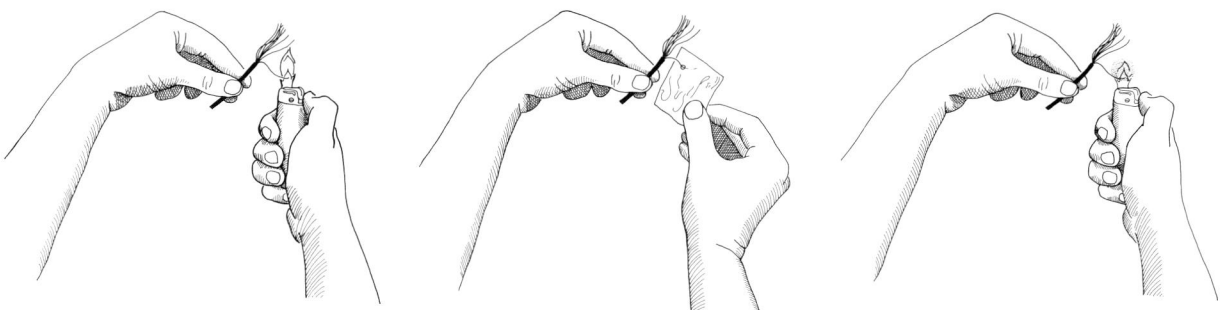

Fig. 10 Beilstein test using copper wire and a flame

small amounts of the sample in containers that facilitate manipulation and viewing of the sample and reaction. The technique generally involves sealing the soft glass of a Pasteur pipette below the taper with flame heat to create a funnel-like trap for a small sample. The sample is placed inside the pipette in or near the trap. Depending on the type of test, a small piece of pH indicator paper or a piece of reagent-impregnated spot-test paper may be folded and lodged at the wide end of the pipette which is then sealed (a laboratory sealing wrap may be used). As the sample is heated over the flame, it may sublime, melt or decompose with characteristic color change, or a gas may be evolved. Recognizable details may be observed within the pipette chamber as evolved gases reach the test papers. A specific odor may also be detected after the pipette is unsealed (see Fig. 11). A wooden clothes pin is a handy tool for holding the pipette during tests.

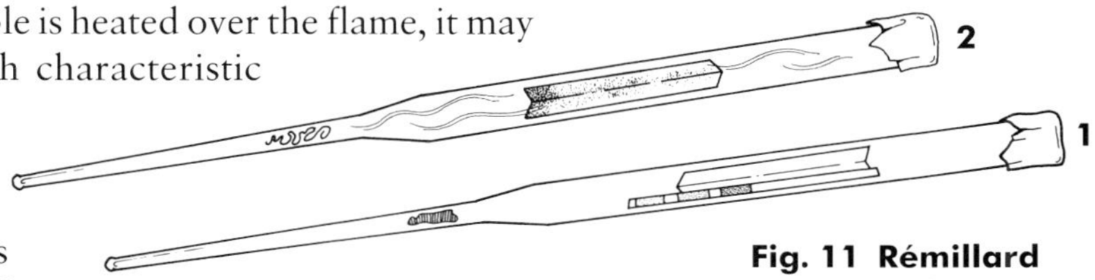

Fig. 11 Rémillard pipette technique

3.2.5 Electrolysis sampling set-up

Many spot tests for metals require that the sample be taken into solution. To minimize the damage caused by the dissolution of metal on an artefact surface, the use of an electrolysis device is recommended. This hand-made device consists of a 6 volt battery, two lengths (50cm each) of 16 gauge insulated copper electrical wire, and four alligator clips. In using this method, the sample of metal object material to be characterized is obtained by dissolving it electrolytically and transferring it onto a piece of filter paper. The metal object is connected into an electrical circuit at the positive pole. A piece of filter paper dampened with an ionic solution (i.e. tap water or NaCl solution) may be used to pad the surface of a delicate object from the pressure of the alligator clip. The circuit is completed by placing the small piece of reagent-treated paper in a forceps or tweezers (that have direct contact with the wet zone of the paper) which is then connected to an alligator clip and wire at the negative pole. Close proximity of the electrodes may facilitate the reaction. The effect is a dissolution of anode metal by electrolysis and the cations that are produced are transferred to the surface of the reagent treated paper that is in contact with only a very small area of artefact material. The use of graphite rods (mechanical pencil leads) to further reduce the area of contact has been successfully tried. The graphite rods,

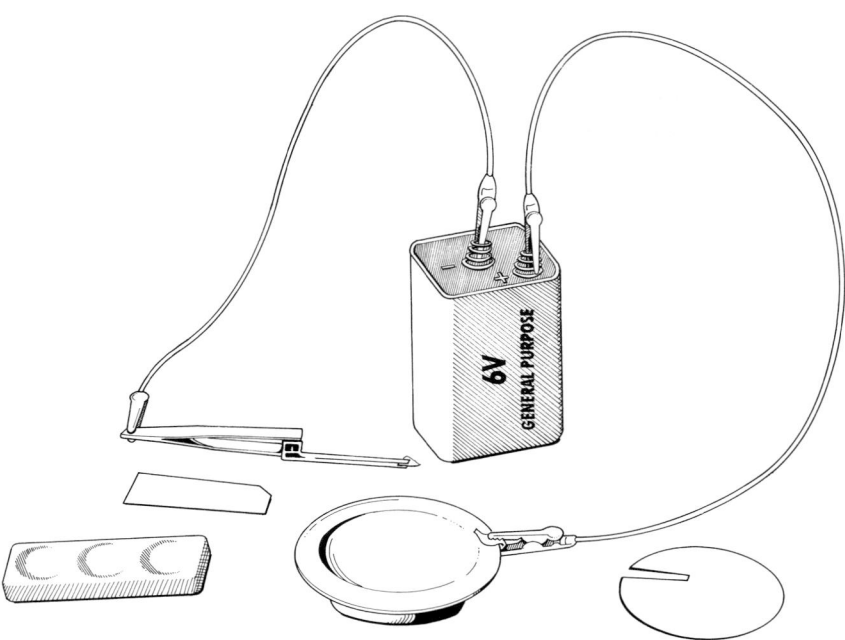

Fig. 12 Electrolysis set-up

which conduct electrical current, are held by the alligator clips. The circuit is completed by touching the reagent-activated test paper that is held to the object with the graphite rod. Special caution is advised when testing metallic threads using this method as the current could cause the threads to burn through to the textile (see Fig. 12).

3.2.6 Centrifuge

A centrifuge is an instrument with a chamber that revolves at high speed to impart a centrifugal force that separates a mixture of two substances of different density. Material of higher density is thrown towards the outer portion of the chamber (bottom of the tube) and that of lower density is concentrated at or near the inner portion (top of the tube). Centrifuges are used in spot testing to quickly concentrate a precipitate into a small volume so that clear solution may be obtained efficiently and corrosive liquids can be manipulated more easily. The micro centrifuge or Capsulefuge™ is particularly useful because minute quantities of precipitate and supernatant can be isolated. The correct size, quantity and placement of microtubes or capsules is essential for stable machine function. Breakage, spillage and contamination of the tests are also prevented when the procedures are followed properly (see Fig. 13).

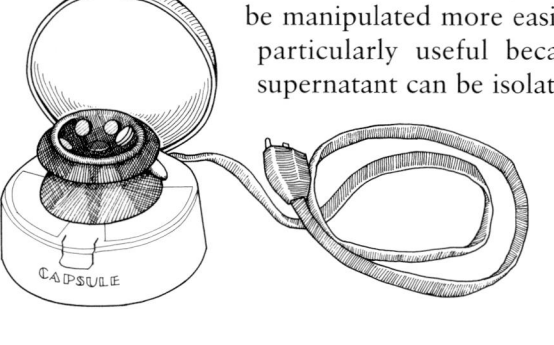

Fig. 13 Micro centrifuge

30 Material characterization tests for objects of art and archaeology

3.2.7 Melting point device

The term melting point refers to the temperature at which a solid substance begins to change into a liquid. Pure organic compounds have sharp melting points; contaminants tend to reduce the melting point. An electric melting point apparatus or Mel-temp™ device works by housing the sample and a thermometer in a metal block. An electric current is used to heat the block. A thermometer (alcohol) or thermocouple and capillary tubes of known and unknown sample are inserted into holders on the hot metal block. The temperature at which the solid melts is observed through a small eyepiece. The point or range when melting occurs can be specific for certain materials and even though an identification may not be possible, some level of characterization may be. Capillary tubes should be filled with a minimum amount of sample. Used tubes should be discarded as glass waste (see Fig. 14).

3.3 Conclusion

There are many advantages to following scientific methodology and using experimental design to conduct material characterization tests. When a test performs well there is usually a good color reaction and it appears quickly. Repetition in testing builds confidence in spot-testing technology. The more a scientific methodology is followed, the greater the success and the more confidence will be gained. Confidence in the test results may be critical to a discussion related to research, collection management, or conservation treatment decisions. Classical spot-test analyses are elegant in their simplicity, however frustration is bound to occur if the results are irregular due to poor experimental design. Refining spot-test techniques to achieve useful results with the extremely small and impure samples from artefacts is challenging. Much more research and development can be done.

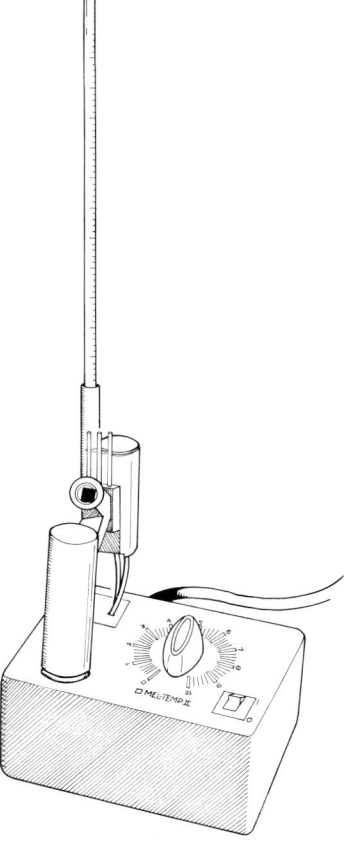

Fig. 14 Mel-temp™ **melting point device**

4 Spot tests for metals

Aluminum	34
Antimony	38
Arsenic	40
Chromium	44
Copper	46
Gold	54
Iron	62
Lead	66
Mercury	72
Nickel	78
Silver	82
Tin	88
Zinc	92

Test for aluminum ions using aluminon (aurintricarboxylic acid)

Purpose
The determination of the presence of aluminum (Al^{3+}) ions. This test can be used to identify the presence of alum sizing on paper, alum used in tanning leather, and aluminum silicates such as kaolin, which is sometimes used as a white pigment, ground, or surface whitener. The test does not work very well with aluminum metal.

Principle
Aluminum (Al^{3+}) ions form a red lake (red precipitate) with aluminon.

$$Al^{3+} (aq) + C_{22}H_{23}N_3O_9 (aq) \rightarrow \text{red complex } (s)$$
$$\text{(aluminon)}$$

Equipment

- balance (to weigh 0.1g)
- graduated cylinder
- acid-resistant reagent containers
- stainless steel tweezers
- dropper
- pH-indicator paper
- filter paper
- microspatula
- spot-test plate

Reagents and safety

- **Aluminon** [aurintricarboxylic acid, ($C_{22}H_{23}N_3O_9$)]: irritant; slight health rating
- **Ammonium acetate** ($NH_4CH_3 \cdot COO$): irritant; slight health rating
- **Concentrated ammonium hydroxide** (NH_4OH)[household ammonia will do]: toxic, irritant, and corrosive; severe health rating
- **Hydrochloric acid** (HCl): toxic and corrosive; severe health rating
- **Water** [distilled]

Protection
Wear goggles, gloves and protective clothing when handling HCL and NH_4OH.

Other tests to consider or confirm results

- Test for aluminum metal using Alizarin Red S and electrolysis (p. 36).

Reference
Feigl, Fritz, & Vinzenz Anger. 1972. *Spot Tests in Inorganic Analysis*, 6th English edition. New York: Elsevier.

Reagent preparation

- **Aluminon solution**: add 0.03g of aluminon to 30mL distilled water
- **Ammonium acetate solution**: add 0.3g of ammonium acetate to 30mL distilled water
- **0.1M hydrochloric acid (HCl) (1:115):** add 0.3mL conc. HCl to 35mL distilled water. *Alternative method:* add one drop of conc. HCl to 10mL of distilled water (in this situation the exact concentration is not critical). **ALWAYS ADD ACID TO WATER!**

Method of sampling

The test may be carried out directly on the object but never on a paper object. Where possible, some material may be removed for testing on the spot-test plate. The acid will etch the artefact so care should be taken to select an inconspicuous area and the droplet of chemical should be removed, the area swabbed with distilled water as soon as possible and the artefact dried.

Procedure

1. **Paper or leather**   The aluminon solution can be placed directly on an inconspicuous spot or on the sample taken and observed for the formation of a red or pink color.
   Kaolin   Place a few grains on a piece of filter paper dampened with distilled water and then place a drop of the aluminon solution on them. Observe under magnification for the formation of a red or pink color on the grains.
   Metal artefacts   Place a tiny drop of the HCl solution on an inconspicuous place on the object or, a few drops of HCl can be placed in a test tube with some of the sample. Wait approximately 10 minutes for the aluminum ions to go into solution.
2. Place a drop of aluminon solution in a glass dish or on a microscope slide.
3. Add a drop of ammonium acetate solution.
4. Measure the pH of this solution with pH-indicator paper. It should be slightly basic; if not, add another drop of ammonium acetate until the solution becomes slightly basic. The drop may lose some of its pink color.
5. Wick some of the acid from the object or sample test tube onto a small piece of filter paper held with tweezers.
6. Dip the filter paper into the aluminon/ammonium acetate solution.
7. Expose the filter paper for a few minutes to ammonium hydroxide vapor by holding it over the mouth of a container of conc. ammonium hydroxide or over a shallow dish of less concentrated ammonium hydroxide e.g. household ammonia. White vapors will be evolved from the filter paper as the acid is being neutralized by the ammonia vapor.
8. Swab the artefact with distilled water to remove the chemicals, then dry.
9. Allow the filter paper to dry.

Observations and interpretation

The formation of a red or pink precipitate indicates the presence of aluminum ions. The red spot darkens with time. If the material does not contain aluminum ions the color will fade with time.

Storage and reagent shelf life

Hydrochloric acid is stable and should be stored in a sealed, acid-resistant container in ambient conditions. Ammonium hydroxide, ammonium acetate and aluminon are stable when stored in sealed containers in ambient conditions.

Test for aluminum metal using Alizarin Red S and electrolysis

Purpose
To test metal objects to determine if they contain aluminum.

Principle
Small amounts of metal are dissolved by electrolysis onto papers soaked in hydrochloric acid (HCl) and then tested with Alizarin Red S. By using pointed paper test strips the area exposed to electrolysis can be greatly reduced.

$$Al^{3+} (aq) + C_{14}H_7NaO_7S \cdot H_2O\ (aq) \rightarrow \text{pink-red complex}$$
$$\text{(Alizarin Red S)}$$

Equipment

- electrolysis cables
- balance (to weigh 0.01g)
- filter paper
- acid-resistant reagent container
- 6V battery
- cotton swabs
- graduated cylinder
- spot-test plate
- droppers
- microspatula
- stainless steel tweezers

Reagents and safety

- **Acetone**: flammable and irritant; slight health rating
- **Alizarin Red S** [sodium alizarin sulphonate]: irritant; moderate health rating
- **Hydrochloric acid** (HCl): toxic and corrosive; severe health rating
- **Water** [distilled][tap]

Protection
Wear goggles, gloves and protective clothing when handling HCl.

Other tests to consider or confirm results

- Test for aluminum ions using aluminon (aurintricarboxylic acid) (p. 34).
- Test to determine specific gravity using an electronic analytical balance (p. 190).

Reference
Vogel, Arthur I., & G. Svehla. 1996. *Vogel's Qualitative Inorganic Analysis*, revised. Harlow, England: Longman.

Reagent preparation

- **0.1M hydrochloric acid (HCl) solution (1:115)**: add 0.3mL of concentrated HCl to 35mL distilled water or add one drop of concentrated HCl to 10mL of distilled water (in this situation the exact concentration is not critical). **ALWAYS ADD ACID TO WATER!**
- **0.1% Alizarin Red S solution**: dissolve 10mg (0.01g) of Alizarin Red S in 10mL of distilled water.

Method of sampling

The test may be carried out directly on the object but, where possible, some material may be removed for testing on the spot-test plate. The acid will etch the artefact so care should be taken to select an inconspicuous area and the droplet of chemical should be removed, the area swabbed with distilled water as soon as possible, and the artefact dried.

Procedure

1. Degrease an inconspicuous area of the object with acetone on a cotton swab. This is an important step, which should not be overlooked – the reaction will not take place unless there is good contact with the metal.
2. Connect the electrolysis cables. Attach one clip to the positive pole of the 6-volt battery and the other end to the object. If the metal is soft, the alligator clip may scratch the object. To avoid this, filter paper soaked in tap water can be used as a pad for the alligator clip. If this method is used, the filter paper must always be wet to conduct the current.
3. Place a few drops of the HCl solution on the spot-test plate.
4. Using a pair of stainless steel tweezers, dip a small piece of filter paper into the HCl solution.
5. Attach the tweezers to the negative pole of the 6-volt battery with the other length of wire using the remaining alligator clips. It is important that the tweezers hold the filter paper well within the wet part so that current will be conducted by the tweezers to the filter paper and on through the object.
6. Hold the tip of the filter paper to the object for 10 seconds.
7. Remove the filter paper from the object.
8. Wick 1 drop of Alizarin Red S solution to the filter paper.
9. Swab the artefact with distilled water to remove the chemicals, then dry.

Observations and interpretation

It is a good idea to run this test with a known aluminum source (e.g. aluminum foil) first to observe what color is formed. The formation of a light red or pink color indicates the presence of aluminum in the object. Other metals will give reddish purple and purple colors. The positive for tin and lead is a little deeper red than that for aluminum; more blood red in hue. The tin positive is a deeper red than an aluminum positive, with some hint of purple.

Storage and reagent shelf life

Hydrochloric acid is stable when stored in a sealed, acid-resistant container in ambient conditions. Alizarin Red S solution is stable in ambient conditions for about a month. Sodium chloride is stable in ambient conditions.

Test for antimony using spot-test papers

Purpose
To determine if antimony is present in a metal object (e.g. bronze).

Principle
A piece of treated test paper is activated with dilute HCl and exposed to the object. A color change occurs if the object contains antimony.

$$Sb^{3+} + \text{test paper} \rightarrow \text{pink complex}$$

Equipment
- cotton swabs
- stainless steel tweezers
- acid-resistant containers

Reagents and safety
- **Antimony test papers** [Macherey-Nagel]; no health rating
- **Acetone**: flammable and an irritant; slight health rating
- **Hydrochloric acid** (HCl): is toxic and corrosive; severe health rating
- **Water** [distilled]

Protection
Wear goggles, gloves and protective clothing when handling HCl.

Other tests to consider or confirm results
None.

Reference
Laver, Marilyn. 1978. Spot Tests in Conservation: Metals and Alloys. *International Committee of Museums (ICOM), Committee for Conservation,* 5th Triennial Meeting, Zagreb, 78/23/8:1–11, preprint.

Reagent preparation

- **0.1M hydrochloric acid (HCl) (1:115)**: Add 1mL of concentrated HCl to 115mL of distilled water. *Alternative method:* add two drops of concentrated HCl to 10mL of distilled water (in this situation the exact concentration is not critical). **ALWAYS ADD ACID TO WATER!**

Method of sampling

Tests are performed directly on the object. The acid will etch the surface of the artefact, so the test should be carried out on an inconspicuous area. The acid should be washed off as soon as possible by swabbing with distilled water and the artefact should then be dried.

Procedure

1. Degrease the area to be tested with cotton swabs and acetone. This is an important step that should not be overlooked. The reaction will not take place unless there is good contact with the metal.
2. Dip a piece of antimony test paper into the 0.1M HCl and place it on the surface of the object for a minute or two.
3. Clean the area where acid was in contact with cotton swabs and distilled water and then dry.

Observations and interpretation

Pure antimony will turn the test paper pink in about a minute. Alloys, such as some bronzes, that contain small amounts of antimony will take longer (about 2–4 minutes). Tests performed on antimony bronzes indicated that the test papers were able to detect antimony in the alloy at levels as low as 0.18%. If more concentrated HCl is used (e.g. 1:7 dilution) the reaction will be faster and therefore more care must be taken not to damage the artefact.

Note

The paper turns yellow when the drop of acid is added.

Storage and reagent shelf life

Hydrochloric acid is stable when stored in a sealed, acid-resistant container in ambient conditions. The test papers are stable in ambient conditions.

Test for arsenic compounds using spot-test papers

Purpose
To determine the presence of arsenic compounds on wallpaper, natural history specimens or ethnographic artefacts that have been treated with arsenic used as a pesticide. This test may also be used to identify the arsenic component in arsenical bronzes.

Principle
Hydrogen gas (H_2) is formed by reaction of zinc metal (Zn) with hydrochloric acid (HCl). Arsenic compounds, if present in the solution, are reduced to arsine gas (AsH_3) by reacting with nascent hydrogen. The arsine gas reacts with the mercury(II) bromide-treated paper ($HgBr_2$), forming colored compounds.

$$Zn\ (s) + 2HCl\ (aq) \rightarrow 2H^{\bullet}\ (g) + ZnCl_2\ (aq)$$

$$As^{3+}\ (aq) + 3H^{\bullet}\ (g) \rightarrow AsH_3\ (g)$$

$$AsH_3\ (g) + HgBr_2\ (aq) \rightarrow As(HgBr)_3\ (aq) + HBr$$

(yellow to brown)

Equipment
- Erlenmeyer flask (200mL)
- microspatula
- cotton swabs
- large glass test tubes (20mL)
- laboratory wrapping film
- scissors

Reagents and safety
- **Arsenic test papers** [Macherey-Nagel]: toxic
- **Hydrochloric acid** (HCl): toxic and corrosive; severe health rating
- **Water** [distilled]
- **Zinc metal (Zn) filings or powder.** *Note:* Zinc metal filings (Analan) or powder that do not contain arsenic as an impurity must be used

Protection
- Handle items contaminated with arsenic with gloves, and wear a dust mask.
- Wear goggles, gloves and protective clothing when handling HCl.
- The arsine gas evolved during the test is very toxic and should not be inhaled. A fume hood should be used for this test.
- If the material remaining in the reaction vessel tested positive for arsenic then it should be rinsed with tap water into a properly labeled container suitable for storage and should then be disposed of as hazardous waste. A small amount of isopropyl alcohol should be used for final rinsing; this will aid drying of the reaction vessel.

Other tests to consider or confirm results
Arsenic testing kits are available from: E M Science, a division of E M Industries, VWR Scientific, Thomas Scientific (Merckoquant 10026 Arsenic test, Cat. No. 3108-L55).

- To detect arsenic in bronzes: Test for arsenic or phosphorus in copper metal using iron(III) chloride (p. 42).

Reference
Knapp, Anthony M. 1993. Arsenic Health and Safety Update. *Conserv-O-Gram* No. 2/3. Harpers Ferry, WV: Curatorial Services Division, National Park Service, U.S. Department of the Interior.

Reagent preparation

All reagents can be purchased ready for use.

Method of sampling

The object can be sampled by either removing obvious crystalline or powder residues, or by rolling fine cotton swabs dampened with distilled water on several areas of the specimen, especially in cracks or crevices. To test arsenical bronzes, a small sample (5mg or more) of filings or drillings from the metal to be tested or some of the corrosion products may be placed in a test tube with a few drops of (concentrated) HCl to dissolve them. Note, the corrosion products of arsenical bronzes may be free of arsenic.

Procedure

1. If swabs were used to remove the sample, the cotton tips should be broken off and placed in an Erlenmeyer flask with 25mL of distilled water. After an hour, place 5mL of this solution in a test tube.
2. Add 10 drops of concentrated HCl (if bronze is being tested there is no need to add more HCl).
3. The test tube must have a closure through which the test paper can be inserted. If no closure is available one can be fashioned from a small piece of laboratory wrapping film (Parafilm™). Cover the top of the test tube with a one inch square of laboratory wrapping film (Parafilm™) and then fold it down along the sides. A small slit can be cut in the top for the test paper to be inserted. The closure should be loose fitting (not airtight) otherwise pressure will build up and there will be a danger of explosion.
4. Insert the test paper through the slit in the test tube closure. About 1cm of the test paper should be left projecting from the top of the test tube. It should be folded over to prevent the test paper from falling into the test tube. The test papers from Macherey-Nagel are rather wide so cut them lengthwise to make them easier to fit in the test tube; this is also more economical.
5. Place a scoop full (from the microscoop or spatula) of zinc powder in the test tube.
6. Quickly replace the laboratory wrapping film (Parafilm™) cap and the test paper.
7. Allow to stand for 30 minutes.

Observations and interpretation

If arsenic is present the paper will turn yellow or brown depending on the concentration of arsenic present. Small spots of yellow or brown also signify a positive for arsenic as they indicate the formation of mixed arsenic mercury halogenides ($AsHgBr_2$).

Storage and reagent shelf life

Hydrochloric acid is stable when stored in a sealed, acid-resistant container in ambient conditions. Zinc metal granules are stable in ambient conditions, but can react violently with strong oxidizing agents, strong acids and strong alkalis. The stability of arsenic test papers is unknown but the instructions state that the test kit should be stored in a refrigerator.

Test for arsenic or phosphorus in copper metal using iron(III) chloride

Purpose
To determine the presence or traces of arsenic or phosphorus in copper alloys.

Principle
The formation of a dark spot after treatment with iron(III) chloride ($FeCl_3$) indicates the presence of arsenic or phosphorus in the copper alloy. This test can detect arsenic content at levels as low as 0.07% and phosphorus content at levels as low as 0.09% in the copper alloy.

$$\text{soluble arsenates} + FeCl_3\ (aq) \xrightarrow{acid} \text{arsenic-iron complex (dark spot)}$$

Equipment
- balance (to weigh 0.1g)
- droppers
- graduated cylinder
- acid-resistant reagent container

Reagents and safety
- **Iron(III) chloride** (ferric chloride, $FeCl_3$): corrosive; moderate health rating
- **Hydrochloric acid** (HCl): toxic and corrosive; severe health rating
- **Water** [distilled]

Protection
Wear goggles, gloves and protective clothing when handling HCl.

Other tests to consider or confirm results
- Test for arsenic compounds using spot-test papers (p. 40).

Reference
International Nickel Company, Inc. 1951. *Rapid Identification (spot testing) of Some Metals and Alloys*. New York, NY: Development and Research Division, International Nickel Company.

Reagent preparation

- **Acidified iron(III) chloride (FeCl$_3$) solution**: 1g FeCl$_3$ is dissolved in 1mL of concentrated HCl and then added to 9mL of distilled water. ALWAYS ADD ACID TO WATER!

Method of sampling

Tests are performed directly on the object. The FeCl$_3$ solution will leave a mark on the metal, so an inconspicuous place on the object should be selected for testing and the droplet of chemical should be removed, the area swabbed with distilled water as soon as possible, and the artefact dried.

Procedure

1. Place one drop of the FeCl$_3$ solution on an inconspicuous area of the object.
2. Leave for 15–30 seconds.
3. Rinse with distilled water to remove the chemicals and then dry.

Observations and interpretation

The formation of a dark spot indicates the presence of arsenic or phosphorus. The dark spot forms immediately.

Storage and reagent shelf life

Hydrochloric acid is stable when stored in a sealed, acid-resistant container in ambient conditions. The FeCl$_3$ solution is stable when stored in a sealed, acid-resistant container in ambient conditions.

Test for chromium using diphenylcarbazide and electrolysis

Purpose
To determine the presence of chromium in metal objects, tanned leather, or pigments.

Principle
Small amounts of chromium are dissolved by electrolysis and then tested with a solution of diphenylcarbazide. During the reaction chromate is reduced to chromium(III), and diphenylcarbazide is oxidized to diphenylcarbazone. These reaction products produce a characteristic blue-violet colored complex. By using pointed paper test strips, the area exposed to electrolysis can be greatly reduced. To test materials such as pigments or leather samples it is necessary to dissolve some of the material in concentrated nitric acid (see p. 30).

$$Cr\,(s) \xrightarrow{acid} Cr^{6+}\,(aq)$$

$$Cr^{6+}\,(aq) + \underset{\text{diphenylcarbazide}}{C_{13}H_{14}N_4O} \rightarrow \underset{\text{(blue-violet)}}{(C_{13}H_{10}N_4O)^{3+}}$$

Equipment

- electrolysis cables
- balance (to weigh 0.01g)
- filter paper
- reagent containers
- 6V battery
- cotton swabs
- graduated cylinder
- spot-test plate
- droppers
- microspatula
- stainless steel tweezers

Reagents and safety

- **Acetone:** flammable and an irritant; slight health rating
- **Diphenylcarbazide:** irritant; slight health rating
- **Nitric acid (HNO_3):** corrosive, oxidizing, and toxic; severe health rating
- **Ethanol** [ethyl alcohol]: flammable and an irritant; slight health rating
- **Water** [distilled][tap]

Protection
Wear goggles, gloves and protective clothing when handling HNO_3.

Other tests to consider or confirm results
None.

Reference
Vogel, Arthur I., and G. Svehla. 1996. *Vogel's Qualitative Inorganic Analysis*, revised. Harlow, England: Longman.

Reagent preparation
- **0.5M nitric acid (HNO_3) solution (1:30):** add 1mL concentrated HNO_3 to 30mL distilled water. ALWAYS ADD ACID TO WATER!
- **Diphenylcarbazide solution:** add 0.01g diphenylcarbazide to 10mL ethanol.

Method of sampling
Tests are performed directly on the surface of the object. The acid will etch the artefact so care should be taken to select an inconspicuous area and the droplet of chemical should be removed, the area swabbed with distilled water as soon as possible, and the artefact dried. For pigments and leathers a small sample must be removed for testing.

Procedure
For metal objects
1. Degrease a test area on the object with acetone and swabs. This is an important step and should not be overlooked. The reaction will not take place unless there is good contact with the metal. It may also be necessary to degrease the area where the clip is attached.
2. Connect the electrolysis cables, attach one clip to the positive pole of the 6-volt battery and the other end to the object. If the metal is soft, the alligator clip may scratch the object. To avoid this, filter paper soaked in tap water can be used as a pad for the alligator clip. If this method is used, the filter paper must always be wet to conduct the current.
3. Place a few drops of the HNO_3 solution onto the spot-test plate.
4. Using a pair of stainless steel tweezers, dip a small piece of filter paper into the HNO_3 solution.
5. Attach the tweezers to the negative pole of the 6-volt battery with the other length of wire using the remaining alligator clips. It is important that the tweezers hold the filter paper well within the wet part so that current will be conducted by the tweezers to the filter paper and on through the object.
6. Hold the tip of the filter paper to the object for about 5 seconds (you really don't need much longer).
7. Remove the filter paper and clip from the object, rinse both areas by swabbing with distilled water then dry.
8. Add 1 drop of the diphenylcarbazide solution to the filter paper.

For pigments and leather
1. *Pigments:* place a small sample of pigment on a piece of filter paper.
   Leather: place a small sample in a spot-test plate.
2. Place a drop of concentrated nitric acid on the sample and wait about a minute.
3. *Leather:* transfer a drop from the spot-test plate to a filter paper.
4. Add a drop of diphenylcarbazide solution to the filter paper.

Observations and interpretation
The formation of a violet color indicates the presence of chromium. It is important to note that the color you are looking for is a bright violet, and not red or a reddish purple. Other white metals such as silver, nickel, tin, iron, zinc, and aluminum will give reddish and purplish colors. It is best to perform this test on a known chromed object to see the color formed before testing other white metals so that they may be differentiated. This will help to confirm the presence of chromium in the sample. Also, after electrolysis the base metal of the chromed object will affect the color but the color produced after the addition of diphenylcarbazide will be the same. Stainless steel contains chrome, which will give a positive.

Storage and reagent shelf life
Nitric acid is stable when stored in a sealed, acid-resistant container in ambient conditions. The diphenylcarbazide solution should be made freshly before each use.

Test for copper using spot-test papers

Purpose
To identify the presence of copper in metallic objects or copper contained in surface oxidation products or pigments.

Principle
A specially treated paper (Cuprotesmo) is moistened with de-ionized water and applied to the surface of an object. In the case of pigments or corrosion products, some of the sample is placed on the moistened paper. The paper changes color if copper (Cu^{2+}) ions are present.

$$Cu^{2+} \,(aq) + \text{test paper} \rightarrow \text{pink-purple complex}$$

Equipment
- dropper
- magnification (may be necessary for pigments)

Reagents and safety
- **Cuprotesmo test papers** [Macherey-Nagel]; no health rating
- **Water** [de-ionized]

Protection
None.

Other tests to consider or confirm results
- Test for copper using nitric acid and ammonia (p. 48).
- Test for copper-based pigments using potassium ferrocyanide (p. 50).
- Test for copper using rubeanic acid (p. 52).

Reference
Laver, Marilyn. 1978. Spot Tests in Conservation: Metals and Alloys. *International Committee of Museums (ICOM), Committee for Conservation*, 5th Triennial Meeting, Zagreb, 78/23/8:1–11, preprint.

Reagent preparation
None.

Method of sampling
Tests are performed directly on the surface of the object, or the test paper can be wetted and pigments placed directly on the paper. Do not touch the papers directly with your hands. The test paper may leave a mark on the metal where it was in contact. It is important to carry out the test on an inconspicuous area and to remove the test paper as soon as a positive result is noted.

Procedure

1. Wet a piece of the Cuprotesmo test paper by wicking up a drop of de-ionized water.
2. Apply the wetted test paper to the object.
3. Rinse the test area with distilled or de-ionized water and then dry.

Alternative procedure

1. Wet a piece of the Cuprotesmo test paper by wicking with a drop of de-ionized water.
2. Hold the paper beneath the pigmented area and cause a few grains of pigment to fall on the test paper.

Observations and interpretation
The paper will turn pink/purple if copper ions are present. The test papers work better if corrosion is present because the copper ions are in more soluble form. If there are no corrosion products, the test papers will take longer to change color. The test papers sometimes turn yellow when other metals are present. When pigments are tested, small spots of pink color will form around the pigment. It may be necessary to examine the paper under low power magnification. The use of de-ionized water is preferred over distilled water as metal ions must be attracted to the test paper for a reaction. Distilled water is not as ion hungry as de-ionized water and will therefore take much longer to react.

Storage and reagent shelf life
The test papers are stable in ambient conditions.

Test for copper using nitric acid and ammonia

Purpose
To determine if a metal alloy or pigment contains copper.

Principle
The dissolution of a trace amount of the metal with nitric acid (HNO_3), which is then exposed to ammonia vapor from a concentrated solution of ammonia (NH_4OH), which forms a copper-amine complex, which turns blue.

$$Cu^{2+} \, (aq) + 6NH_3 \, (g) \rightarrow [Cu(NH_3)_6]^{2+} \, (aq)$$
$$\text{(blue)}$$

Equipment

- dropper
- graduated cylinder
- tweezers
- filter paper
- acid-resistant reagent container

Reagents and safety

- **Concentrated ammonia solution (NH_4OH)** [or household ammonia]: toxic, irritant and corrosive; severe health rating
- **Nitric acid (HNO_3)**: corrosive, oxidizing and toxic; severe health rating
- **Water** [distilled]

Protection
Wear goggles, gloves and protective clothing when handling HNO_3 and NH_4OH. The reaction should be performed in a fume hood.

Other tests to consider or confirm results

- Test for copper using spot-test papers (p. 46).
- Test for copper using rubeanic acid (p. 52).
- Test for copper-based pigments using potassium ferrocyanide (p. 50).

Reference
Moss A. A. 1956. *The Identification of Metals. Handbook for Museum Curators*, Part B, *Museum Technique*, Section 8, 2–8. London: Museums Association.

Reagent preparation
- **8M nitric acid (HNO$_3$) solution (1:1)**: Carefully add 15 mL of analytical grade, concentrated nitric acid to 15 mL of distilled water. **ALWAYS ADD ACID TO WATER!**

Method of sampling
The test is performed directly on the surface of the object. The acid will etch the object therefore it is important to choose an inconspicuous area for testing to remove the acid as soon as observations are made.

Procedure
1. Place a tiny drop of the HNO$_3$ solution on an inconspicuous area on the object.
2. Look for effervescence and dissolution of the surface.
3. After about 30 seconds, wick the liquid on to a small piece of filter paper and, with tweezers, hold the paper over the mouth of an open container of concentrated ammonia solution. *Note:* Do not perform this step over the bottle of stock solution, because the paper might fall in and contaminate the entire bottle. Pour off a small amount for use in the test.
4. Swab the area with distilled water and then dry.

Alternative procedure for pigments
1. Collect a tiny amount of pigment on a damp piece of filter paper.
2. Place a drop of the HNO$_3$ solution on the pigment particles.
3. With tweezers, hold the filter paper over the mouth of an open container of concentrated ammonia solution.

Observations and interpretation
Copper or copper-containing alloys will dissolve in the dilute nitric acid forming copper nitrate. The copper nitrate, when exposed to ammonia vapor turns blue, or blue-green. If the filter paper turns blue the artefact contains copper.

Sometimes white vapors will be evolved from the filter paper as the acid is being neutralized by the ammonia vapors because NH$_4$NO$_3$ (a salt) is being formed.

Storage and reagent shelf life
Nitric acid (HNO$_3$) is stable when stored in a sealed, acid-resistant container in ambient conditions. Nitric acid may turn yellow with time, but that will not affect this test. Ammonium hydroxide is stable when stored in a sealed container in ambient conditions.

Test for copper using potassium ferrocyanide

Purpose
To determine the presence of copper, copper-based pigments, and corrosion products [copper (Cu^{2+}) ions].

Principle
Copper ions in the sample are dissolved and converted to cupric ferrocyanide.

$$2Cu^{2+}\ (aq) + K_4Fe(CN)_6\ (aq) \rightarrow Cu_2Fe(CN)_6\ (s)$$

(red-brown)

Equipment

- balance (to weigh 0.1g)
- graduated cylinder
- scalpel
- droppers
- microspatula
- source of heat (lab oven or infrared lamp)
- filter paper
- acid-resistant reagent containers
- spot-test plate

Reagents and safety

- **Concentrated hydrochloric acid** (HCl): toxic and corrosive; severe health rating
- **Potassium ferrocyanide** [potassium hexacyanoferrate(II), $K_4Fe(CN)_6$]: irritant; slight health rating
- **Water** [distilled]

Protection
Wear goggles, gloves and protective clothing when handling HCl.

Other tests to consider or confirm results

- Test for copper using nitric acid and ammonia (p. 48).
- Test for copper using rubeanic acid (p. 52).
- Test for copper using spot-test papers (p. 46).

Reference
Tabasso M. Laurenzi. 1993. *Mural Paintings Conservation Course – Identification of Pigments Lab Notes*, Part I: *Constituent Materials/Execution Techniques*. Rome, Italy: ICCROM.

Reagent preparation

- **3M hydrochloric acid (HCl) (1:3):** add 5mL concentrated HCl to 15mL distilled water. **ALWAYS ADD ACID TO WATER!**
- **Potassium ferrocyanide solution:** dissolve 1g $K_4Fe(CN)_6 \cdot 3H_2O$ in 25mL distilled water.

Method of sampling

Use a few grains of pigment removed from a painting, mural or stains from soil, or a tiny amount of pigment may be brushed onto a piece of filter paper dampened with distilled water. The sample should be as pure as possible so that it contains no contaminants that might give rise to a false color.

Procedure

1. Dissolve some of the sample on a spot-test plate with concentrated HCl or place a drop of concentrated HCl on the pigment on the filter paper. Some of the sample must dissolve, but not necessarily all of it.
2. Dry completely under an infrared lamp or in a lab oven.
3. A drop of 3M HCl solution may assist the reaction with some tests.
4. Add a drop of the potassium ferrocyanide solution.

Observations and interpretation

The formation of a red-brown color (cupric ferrocyanide) indicates the presence of copper. The reaction takes place almost immediately. The presence of iron(III) ions (Fe^{3+}) may interfere with the color interpretation of this test. Iron reacts with the potassium ferrocyanide solution and turns blue. This may mask the color formed by the copper.

Storage and reagent shelf life

Hydrochloric acid is stable when stored in a sealed, acid-resistant container in ambient conditions. The potassium ferrocyanide solution is stable when stored in a sealed container in ambient conditions.

Test for copper using rubeanic acid

Purpose
To determine if a metal alloy, corrosion product, or pigment contains copper.

Principle
The sample is dissolved in hydrochloric acid (HCl), the solution neutralized with ammonia vapor, and treated with rubeanic acid (dithiooxamide). The formation of an olive green spot indicates the presence of copper.

$$Cu^{2+}\ (aq) + 2H_2N-\underset{\underset{S}{\|}}{C}-\underset{\underset{S}{\|}}{C}-NH_2 \rightarrow \left[\begin{array}{c}\text{Cu complex}\end{array}\right]$$

Equipment

- balance (to weigh 0.1g)
- filter paper
- acid-resistant reagent containers
- test tubes
- droppers
- graduated cylinder
- spot-test plate
- tweezers

Reagents and safety

- **Ammonium hydroxide** (NH_4OH)(household ammonia will work): toxic, an irritant, and corrosive; severe health rating
- **Ethanol** (ethyl alcohol): flammable and an irritant; slight health rating
- **Hydrochloric acid** (HCl): toxic and corrosive; severe health rating
- **Rubeanic acid** [dithiooxamide ($C_2H_4N_2S_2$)]: toxic and an irritant; moderate health rating
- **Water** [distilled]

Protection
Wear goggles, gloves and protective clothing when handling HCl and NH_4OH. The reaction should be performed in a fume hood.

Other tests to consider or confirm results

- Test for copper using nitric acid and ammonia (p. 48).
- Test for copper-based pigments using potassium ferrocyanide (p. 50).
- Test for copper using spot-test papers (p. 46).

Reference
Schramm, Hans-Peter. 1995. *Historische Malmaterialien und Ihre Identifizierung*. Stuttgart: Ferdinand Enke Verlag.

Reagent preparation

- **3M hydrochloric acid (HCl) solution (1:3):** add 2mL concentrated HCl to 6mL distilled water. **ALWAYS ADD ACID TO WATER!**
- **0.08M rubeanic acid solution (1%):** add 0.1g rubeanic acid to 10mL ethanol and stir until dissolved.

Method of sampling

The test is performed on a very small amount of sample or directly on the surface of the metal. The acid will etch the artefact so care should be taken to select an inconspicuous area and the droplet of chemical should be removed, the area swabbed with distilled water as soon as possible, and the artefact dried.

Procedure

1. Place a tiny amount of the sample on the spot-test plate.
2. Add two drops of the HCl solution.
3. Holding a small piece of filter paper with tweezers, dip one end in the solution. If the metal surface is being tested, dip the filter paper in the acid solution and then hold it in contact with the object for a few seconds. Swab the artefact with distilled water to remove the chemicals, then dry.
4. Hold the filter paper over the open mouth of a container of concentrated ammonium hydroxide for a few minutes. White fumes will evolve from the filter paper as the acid is neutralized by the ammonia vapor. Hold the filter paper over the ammonia until no more fumes are evolved.
5. Place a drop of the rubeanic acid solution on the filter paper.

Observations and interpretation

The presence of copper is indicated by the formation of an olive green color in the area where the test solution soaked into the filter paper. Silver and mercury, if present, may interfere.

Storage and reagent shelf life

Hydrochloric acid is stable when stored in a sealed, acid-resistant container in ambient conditions. The solution of rubeanic acid should not be kept for more than a few weeks.

Test to determine gold karat using a touchstone

Purpose
To determine the approximate karat (k) of objects made from gold metal.

Principle
The approximate karat of the unknown is determined by comparing its rate of dissolution in various acids to that of known standards. When the unknown dissolves faster than one of the standards but slower than the next higher standard then the karat of the unknown must lie somewhere between the two.

$$14k \text{ gold or lower} + HNO_3 \ (aq) \rightarrow Au^+ \ (aq) \text{ (alloy dissolves)}$$

$$18k \text{ gold or higher} + HNO_3 \ (aq) \rightarrow \text{no reaction (alloy does not dissolve)}$$

$$18k \text{ gold or higher} + \text{aqua regia} \rightarrow AuCl_4^- \ (aq) \text{ (alloy dissolves)}$$

Equipment
- touchstone (a black stone for testing gold, available from jewelers' supply)
- droppers
- acid-resistant reagent container
- graduated cylinder
- set of gold testing needles (needles with various karat purity of gold on the tips, available from jewelers' supply) or gold of various known karat purity (24, 22, 20, 18, 14, 10)

Reagents and safety
- **Hydrochloric acid** (HCl): toxic and corrosive; severe health rating
- **Nitric acid** (HNO_3): corrosive, oxidizing, and toxic; severe health rating
- **Water** [distilled]

Protection
Wear goggles, gloves and protective clothing when handling HCl and HNO_3.

Other tests to consider or confirm results
- Test to determine gold quality using nitric acid (p. 56).
- Test for gold using tin(II) chloride and electrolysis (p. 58).

Reference
Untracht, Oppi. 1982. *Jewelry Concepts and Technology*. Garden City, New York: Doubleday.

Reagent preparation

- **Aqua regia:** start with one part distilled water (5mL is a good amount), add one part (5mL) concentrated nitric acid (HNO_3), then add three parts (15mL) hydrochloric acid (HCl). **ALWAYS ADD ACID TO WATER!** This solution will turn slightly yellow and chlorine gas will be evolved, therefore it should be prepared and carried out in a fume hood or a well-ventilated area.

Method of sampling

The object is used to make streaks on a touchstone. The scratching on the touchstone may leave a noticeable mark on the artefact. It should be carried out using an inconspicuous part of the artefact.

Procedure

1. Make a long streak (*ca*. 2cm) on the touchstone with the sample to be tested (see p. 25).
2. Make a streak (*ca*. 2cm) on one side with a known gold standard of lower karat than the unknown is expected to be and one of higher on the other side of the unknown. Try to make the streaks the same width and density as the one made by the artefact. If possible, apply the same pressure when making all the streaks, and place the streaks fairly close together so that one drop will cover them.
3. Place a drop of the appropriate test solution on the streaks so that it covers the centre of all the streaks equally. If gold of 14k or less is the anticipated karat then only concentrated nitric acid (HNO_3) should be dropped on the streaks, but if gold of 18k or above is the anticipated karat then aqua regia should be used.

Observations and interpretation

It is important to observe closely the streaks under the test solution. The rate of dissolution of the streaks will be determined by the gold content: the lower the gold content, the faster it will dissolve in the test solution. Thus if the unknown dissolves more slowly than the lower karat streak, but faster than the higher one, it lies between the two. It may be necessary to repeat the test several times with different known gold standards to determine the karat of the unknown. It is important to note that the way the streaks are made influences the rate of dissolution. Obviously a heavy streak of 18k gold will take longer to dissolve than a very light streak of 20k gold. Every streak should be made the same way in both pressure and width. Also the other constituents of the gold alloy (usually copper and silver) affect the rate of dissolution. Two 14k gold alloys with different amounts of copper and silver will dissolve at different rates. This test works well but should only be taken as an approximation. It is also important to note that an object made from gold plate or gold fill will dissolve in the test solution at the same rate as an object from the same karat of solid gold. Other yellow alloys such as brass and bronze will also dissolve at a given rate. It is important to know that the object in question is, indeed, gold. The test only tests the surface of the object. *Note for archaeological gold:* according to Moss (1956), if a gold alloy has lain buried a long time, the surface acquires a higher karat than that of the interior (by preferential corrosion of the more base metal in the alloy).

Storage and reagent shelf life

Hydrochloric and nitric acids are stable when stored in sealed, acid-resistant containers in ambient conditions. Aqua regia will evolve chlorine gas (Cl_2) during storage. The water in the above formula for aqua regia allows the mixture to be stored for a somewhat longer period of time, but the mixture should not be kept for more than a few days.

Test to determine gold quality using nitric acid

Purpose
To distinguish higher karat (k) gold (18k and above) from lower karat gold (14k and below) or copper alloys. This test determines the presence of copper in a metal object.

Principle
Gold alloys of 18k and above are not soluble in nitric acid (HNO_3). Gold alloys of 9–14k are soluble in nitric acid and become brown on the spot where the nitric acid is in contact. Gold alloys lower than 9k, such as tumbaga (20% gold or 5k) turn black where the nitric acid is in contact and leave green copper nitrates on the filter paper. Copper alloys such as brass and bronze turn bright green in contact with nitric acid.

$$18k \text{ gold or higher} + HNO_3 \text{ } (aq) \rightarrow \text{no reaction (gold does not dissolve)}$$

$$Cu \text{ } (s) + 2HNO_3 \text{ } (aq) \rightarrow Cu(NO_3)_2 \text{ } (aq)$$
$$\text{(blue-green)}$$

Equipment
- dropper
- filter paper
- spot-test plate
- tweezers

Reagents and safety
- **Concentrated nitric acid** (HNO_3): corrosive, oxidizing, and toxic; severe health rating
- **Water** [distilled]

Protection
Wear goggles, gloves and protective clothing while handling HNO_3.

Other tests to consider or confirm results
- Test for gold plate using aqua regia (p. 60).
- Test for gold using tin(II) chloride and electrolysis (p. 58).
- Test to determine gold karat using a touchstone (p. 54).

Reference
Moss A. A. 1956. *The Identification of Metals. Handbook for Museum Curators*, Part B, *Museum Technique*, Section 8, 2–8. London: Museums Association.

Reagent preparation
None.

Method of sampling
The test is performed directly on the surface of the object. The acid will etch lower karat gold, so care should be taken to select an inconspicuous area and the droplet of chemical should be removed, the area swabbed with distilled water as soon as possible and the artefact dried.

Procedure
1. Place a drop of concentrated nitric acid (HNO_3) on the spot-test plate.
2. Using tweezers, tear off a tiny piece of filter paper, about the size of a pin head.
3. Using tweezers, dip the edge of the filter paper into the drop of concentrated nitric acid (HNO_3).
4. Place the filter paper on an inconspicuous area of the object.
5. After 30–60 seconds note the color of the object and the filter paper.
6. Wash the acid off the object with distilled water and dry.

Observations and interpretation
Gold alloys >18k will not be affected by the acid; those from 9–14k will turn brown where the acid is in contact; those <9k will turn black where the acid is in contact and the filter paper will turn green. Copper alloys that look like gold (bronze and brass) will turn the filter paper bright green.

Storage and reagent shelf life
Nitric acid (HNO_3) is stable when stored in a sealed, acid-resistant container in ambient conditions.

Test for gold using tin(II) chloride and electrolysis

Purpose
To verify the presence of gold in metal objects.

Principle
Picograms of gold are dissolved by electrolysis onto filter paper soaked in tap water. The filter paper is then treated with tin(II) chloride ($SnCl_2$) which causes brownish black spots to appear on the filter paper. The spots are caused by the reaction of colloidal gold with tin hydroxide (Schramm 1995: 174). By using pointed paper test strips the area exposed to electrolysis can be greatly reduced (see p. 30).

$$Au\ (s) \xrightarrow{electrolysis} Au^{3+}\ (aq)$$

$$2Au^{3+}\ (aq) + 4Sn^{2+}\ (aq) + H_2O\ (l) \rightarrow 2Au\ (s) + Sn(OH)_2\ (s) + 3Sn^{4+}\ (aq) + 2H^+\ (aq)$$
$$\text{(brown)}\quad\text{(purple)}$$

Equipment

- electrolysis cables
- balance (to weigh 0.1g)
- filter paper
- acid-resistant reagent containers
- 6V battery
- cotton swabs
- graduated cylinder
- spot-test plate
- droppers
- microspatula
- stainless steel tweezers

Reagents and safety

- **Acetone**: flammable and an irritant; slight health rating
- **Hydrochloric acid (HCl)**: is toxic and corrosive; severe health rating
- **Tin(II) chloride** ($SnCl_2$) [also known as stannous chloride]: irritant; moderate health rating
- **Water** [distilled][tap]

Protection
Wear goggles, gloves and protective clothing when handling HCl.

Other tests to consider or confirm results

- Test to determine gold karat using a touchstone (p. 54).
- Test for gold plating using aqua regia (p. 60).
- Test to determine gold quality using nitric acid (p. 56).

References
Laver, Marilyn. 1978. Spot Tests in Conservation: Metals and Alloys. *International Committee of Museums (ICOM), Committee for Conservation*, 5th Triennial Meeting, Zagreb, 78/23/8:1–11, preprint.

Reagent preparation

- **5.8M hydrochloric acid (HCl) solution (1:1):** add 7.5mL concentrated HCl to 7.5mL distilled water. ALWAYS ADD ACID TO WATER!
- **Tin(II) chloride (SnCl$_2$) solution:** add 5g SnCl$_2$ to 15mL of 5.8M HCl.

Method of sampling

Tests are carried out directly on the surface of the object. The alligator clips can scratch the soft metal and the electrolysis can leave a mark on the metal surface so an inconspicuous spot should be selected for testing. If metallic gold thread is being tested, care should be taken so that the current does not melt the thread.

Procedure

1. Degrease an inconspicuous area on the object with acetone on a cotton swab. This is an important step that should not be overlooked. The reaction will not take place unless there is good contact with the metal. It may also be necessary to degrease the area where the clip is attached.
2. Connect the electrolysis cables. Attach one clip to the positive pole of the 6V battery and the other end to the object. If the metal is soft, the alligator clip may scratch the object. To avoid this, filter paper soaked in tap water can be used as a pad for the alligator clip. If this method is used, the filter paper must always be wet to conduct the current.
3. Place a drop of the 5.8M HCl solution on the spot-test plate.
4. Using a pair of stainless steel tweezers dip a small piece of filter paper into the HCl solution. Be sure to hold the filter paper so that one corner or point comes straight out from the tweezers. The filter paper can be properly shaped by cutting it into a triangle with one long side or by folding a square piece of filter paper into a triangle.
5. Attach the tweezers to the negative pole of the 6V battery with the other length of wire, using the remaining alligator clips. It is important that the tweezers hold the filter paper well within the wet part so that current will be conducted by the tweezers to the filter paper and on through the object.
6. Hold the corner or point of the filter paper to the object for about 15 seconds.
7. Remove the filter paper from the object.
8. Add a drop of the tin(II) chloride solution.
9. Rinse the test area with distilled water.

Observations and interpretation

When the tin(II) chloride solution is added the filter paper will turn black at the tip where the paper touched the gold. During the initial electrolysis, the paper touching the object may turn yellow (for 18k and above) or slightly green (for 14k). Gold alloys of 10k and lower give erratic results (sometimes turning green or black during electrolysis, but not changing color when the tin(II) chloride solution is added). Gold-plated objects only give a weak positive for gold or sometimes no reaction at all. It seems that the base metal interferes with the electrolysis, so if an object appears to be made from a high quality gold but does not turn the paper gold or green during electrolysis then it may be plated. Copper and silver do not interfere with the test.

Storage and shelf life of reagents

Hydrochloric acid is stable when stored in a sealed, acid-resistant container in ambient conditions. Tin (II) chloride is a powerful reducing agent. Store it in a sealed container and avoid moisture and heat.

Test for gold plating using aqua regia

Purpose
To determine if an artefact is gold plated or made of a a gold alloy.

Principle
Gold of 18 karats (k) or higher is not soluble in nitric acid (HNO_3) so plated gold will react in a similar way to solid gold when exposed to nitric acid. They may be differentiated using a mixture of acids called aqua regia, which will dissolve gold (even of high karat). Aqua regia is a 1:3 mixture of nitric acid (HNO_3) and hydrochloric acid (HCl) (for more information about aqua regia see Glossary, p. 222). If the object to be tested is gold plated, the aqua regia (placed on a small piece of filter paper) will dissolve the gold to reveal the base metal underneath. If copper is the base metal the filter paper will turn blue-green because the copper ions react with the nitric acid component of the aqua regia to form copper nitrates.

Once the gold plate has been dissolved away with the aqua regia, if the base metal is not a copper alloy it will almost certainly appear to be a white metal. If the test is performed under magnification, the size of the test area can be greatly reduced.

$$14k \text{ gold or lower} + HNO_3 \ (aq) \rightarrow Au^+ \ (aq) \ (\text{gold dissolves})$$

$$18k \text{ gold or higher} + HNO_3 \ (aq) \rightarrow \text{no reaction (gold does not dissolve)}$$

$$18k \text{ gold or higher} + \text{aqua regia} \rightarrow AuCl_4^- \ (aq) \ (\text{gold dissolves})$$

Equipment

- droppers
- magnification
- tweezers
- filter paper
- acid-resistant reagent container
- graduated cylinder
- spot-test plate

Reagents and safety

- **Concentrated nitric acid** (HNO_3): corrosive, oxidizing, and toxic; severe health rating
- **Concentrated hydrochloric acid** (HCl): toxic and corrosive; severe health rating
- **Water** [distilled]

Protection
Wear goggles, gloves, and protective clothing while handling HCl and HNO_3.

Other tests to consider or confirm results

- Test to determine gold quality using nitric acid (p. 56).
- Test for gold using tin(II) chloride and electrolysis (p. 58).
- Test to determine gold karat using a touchstone (p. 54).

Reference
Untracht, Oppi. 1982. *Jewelry Concepts and Technology*. Garden City, New York: Doubleday.

Reagent preparation

- **Aqua regia**: start with one part distilled water (5mL is a good amount), add 1 part (5mL) concentrated nitric acid (HNO_3), then add 3 parts (15mL) hydrochloric acid (HCl). **ALWAYS ADD ACID TO WATER!** This solution will turn slightly yellow and chlorine gas will be evolved, therefore it should be prepared and used in a fume hood or a well-ventilated area. Store this solution in an acid-resistant storage bottle.

Method of sampling

Tests are performed directly on the surface of the object. The acid will etch lower karat gold and completely dissolve the plating to reveal the base metal below, so the test should be carried out on an inconspicuous area and the acid washed off with distilled water as soon as possible.

Procedure

1. Place a drop of aqua regia on the spot-test plate.
2. Using the tweezers tear off a small piece of filter paper about the size of a pin head and dip a corner of it into the drop of aqua regia.
3. Place the filter paper on an inconspicuous area of the object.
4. After 30–60 seconds remove the filter paper and wash the area where the filter paper was in contact with the object with distilled water and then dry, and let the filter paper dry.
5. Under magnification examine the part of the filter paper that was in contact with the object.

Observations and interpretation

Once it has been determined that the artefact is actually made from gold and not one of the yellow-metal alloys then this test can help to determine if the object is plated or made from solid gold. This test requires a little deductive reasoning to interpret correctly. If the filter paper soaked with aqua regia turns green it indicates that there is a large quantity of copper metal present in the artefact. This can only arise from one of two possibilities: the purity of the gold is 14k or less, or the object is plated/filled and the base metal is a copper alloy. The spot where the filter paper touched should be observed under magnification to see if there is copper metal below the gold. This will confirm that it is plate gold. If the filter paper turns yellow but not green, then the artefact is made from solid gold of 18k or higher. The test, *The determination of gold karat using a touchstone* (p. 54), can help to determine the purity of the gold. If the filter paper does not turn green but a white or silvery metal shows below the gold, the artefact is plated but the base metal is a white metal (silver, zinc, lead, tin) and not a copper alloy.

Storage and reagent shelf life

Hydrochloric and nitric acids are stable when stored in sealed, acid-resistant containers in ambient conditions. Aqua regia will evolve chlorine gas (Cl_2) during storage. The water component in the above formula for aqua regia allows the mixture to be stored for a longer period of time.

Test for iron using potassium ferrocyanide

Purpose
To determine if corrosion products, stains, or pigments (such as iron pigments on white, non-iron slip pottery) contain iron(III) ions (also known as ferric ions or Fe^{3+}).

Principle
Iron(III) ions will dissolve in hydrochloric acid and react with potassium ferrocyanide ($K_4Fe(CN)_6$) to form ferric ferrocyanide ($Fe_4(Fe(CN)_6)_3$), a bright blue complex (Prussian blue).

$$4Fe^{3+}\ (aq) + 3K_4Fe(CN)_6\ (aq) \rightarrow Fe_4(Fe(CN)_6)_3\ (s) + 12K^+\ (aq)$$
$$\text{(blue)}$$

Equipment

- balance (to weigh 0.1g)
- graduated cylinder
- microspatula
- tweezers
- droppers
- heat source (lab oven or infrared lamp)
- acid-resistant reagent container
- filter paper
- magnification
- spot-test plate

Reagents and safety

- **Hydrochloric acid** (HCl): toxic and corrosive; severe health rating
- **Potassium ferrocyanide** ($K_4Fe(CN)_6 \cdot 3\ H_2O$) [also known as potassium hexacyanoferrate]: irritant; slight health rating
- **Water** [distilled]

Protection
Wear goggles, gloves and protective clothing when handling HCl.

Other tests to consider or confirm results

- Test for iron using hydrochloric acid (p. 64).

Reference
Tabasso, M. Laurenzi. 1993. *Mural Paintings Conservation Course "Identification of Pigments" Lab Notes, Part I: Constituent Materials/Execution Techniques*. Rome: ICCROM.

Reagent preparation

- 3M hydrochloric acid (HCl) solution (1:3): add 1mL concentrated HCl to 3mL distilled water. **ALWAYS ADD ACID TO WATER!**
- Potassium ferrocyanide solution: add 1g $K_4Fe(CN)_6 \cdot 3H_2O$ to 25mL water.

Method of sampling

A small amount of sample must be removed for testing.

Procedure

1. Dissolve some of the sample on a spot-test plate with concentrated HCl or place a drop of concentrated HCl on the pigments on filter paper. Some of the sample must dissolve, but not necessarily all of it.
2. Dry completely under an infrared lamp or in a lab oven.
3. *Optional*: the addition of a drop of the 3M HCl solution may assist the reaction with some metal surfaces.
4. Add a drop of the potassium ferrocyanide solution.

Alternative procedure for pigments

1. On a filter paper that has been dampened with distilled water, collect a tiny amount of pigment.
2. Place the filter paper on the spot-test plate.
3. Place a drop of concentrated HCl on the pigment on the filter paper.
4. Dry completely under an infrared lamp or in a lab oven.
5. Add a drop of the 3M HCl solution.
6. Add a drop of potassium ferrocyanide solution.
7. Observe under magnification.

Observations and interpretation

The formation of a blue color (Prussian blue) indicates the presence of iron. Copper ions (if present) may interfere with the interpretation of this test. Potassium ferrocyanide also reacts with copper ions to form a reddish brown color. Aluminum reacts to form a blue color. If both iron and copper are present in quantity, the final solution may appear to be slightly purple. If the test is carried out on filter paper, the color formation may only be visible near the pigment particles under magnification

Storage and reagent shelf life

Hydrochloric acid is stable when stored in a sealed, acid-resistant container in ambient conditions. Potassium ferrocyanide is stable in ambient conditions but the aqueous solution decomposes slowly on standing.

Test for iron using hydrochloric acid

Purpose
To confirm that a metal object, corrosion product or pigment contains iron.

Principle
The reaction of iron oxides with hydrochloric acid will form iron(III) chloride [ferric chloride ($FeCl_3$)] which is yellow.

$$Fe^{3+} (aq) + HCl (aq) \rightarrow FeCl_3 (aq)$$
$$\text{(yellow)}$$

Equipment
- dropper
- spot-test plate
- filter paper
- heat source (lab oven, hot plate or infrared lamp)

Reagents and safety
- **Hydrochloric acid (HCl)**: toxic and corrosive; severe health rating

Protection
Wear goggles, gloves and protective clothing when handling HCl.

Other tests to consider or confirm results
- Test for iron using potassium ferrocyanide (p. 62).

Reference
Moss, A. A. 1956. *The Identification of Metals. Handbook for Museum Curators,* Part B, *Museum Technique*, Section 8, 2–8. London: Museums Association.

Reagent preparation
None.

Method of sampling
If the object is intact, the test is carried out on the object. If only corrosion products are present, a few fragments may be tested. Care should be taken to select an inconspicuous area and the droplet of chemical should be removed, the area swabbed with distilled water as soon as possible and the artefact dried.

Procedure
1. For *corrosion products*, place a few tiny fragments on a spot-test plate.
2. Add one or two drops of concentrated hydrochloric acid (HCl). Warming may be necessary.
3. Allow to stand.
4. If testing *metal*, place a drop of concentrated hydrocholoric acid (HCl) on the object and wick it into a piece of filter paper.
5. Swab the artefact with distilled water to remove the chemicals, then dry.

Observations and interpretation
If the corrosion products dissolve and form a yellow solution in the acid, then they contain iron. Iron metal will show a light yellow when the drop of HCl is wicked into the filter paper. *Note:* Pure, clean polished iron or steel does not react as well as corroded metal or pigments. Also copper metal may appear yellowish in this reaction.

Storage and reagent shelf life
Hydrochloric acid is stable when stored in a sealed, acid-resistant container in ambient conditions.

Test for lead using spot-test papers

Purpose
To determine the presence of lead in metal objects, pigments, glazes and corrosion products.

Principle
The surface of the object is tested with a specially treated paper that changes color if a significant amount of lead is present.

$$Pb^{2+} \ (aq) + \text{test paper} \rightarrow \text{pink complex}$$

Equipment

- droppers
- scissors
- tweezers
- magnification

Reagents and safety

- **Plumbtesmo test papers** [Macherey-Nagel]: irritant; slight health rating.
- **Water** [de-ionized]

Protection
None.

Other tests to consider or confirm results

- Test for lead using potassium iodide and electrolysis (p. 68).
- Test for lead using potassium dichromate (p. 70).

Reference
Laver, Marilyn. 1978. Spot Tests in Conservation: Metals and Alloys. *International Committee of Museums (ICOM), Committee for Conservation*, 5th Triennial Meeting, Zagreb, 78/23/8:1–11, preprint.

Reagent preparation
None.

Method of sampling
Tests are performed directly on the surface of the object. The test papers may leave a pinkish-red mark on the object if left on for a long time.

Procedure

1. Wet a small piece of the Plumbtesmo test paper by wicking in a drop of de-ionized water. DO NOT SATURATE THE PAPER (the reagent may be washed away). Do not touch test papers directly with hands.
2. With tweezers, place the wet test paper on the surface of the object.
3. Apply pressure to the test paper using the bulb end of a plastic pipette if greater surface contact is needed, such as in the case of an irregular or uneven surface.

Alternative procedure for pigments and powdered corrosion products

1. Wet a piece of the Plumbtesmo test paper by wicking in a drop of de-ionized water. The amount of water is critical – too much will wash the reagent away and with too little there will be no reaction.
2. Hold the paper beneath the pigment area and cause a few grains of pigment to fall onto the test paper. It may be necessary to look at the paper under low power magnification.

Observations and interpretation
The paper turns light orange when the drop of water is first placed on it. After about a minute the orange color disappears. The presence of lead is indicated if the paper turns red or pink. The paper should change color quickly in the presence of lead. In the case of trace amounts however, the change may not be evident until the paper dries. The paper is very sensitive and does not give false positives with other metals. The test was successful in detecting lead in bronzes with approximately 9% lead content, pewter with a high lead content, etched areas of leaded glass, and lead glaze on ceramics. However on high quality pewter where the lead content is low, the papers only detected lead in corroded areas. The use of de-ionized water is preferred to distilled water.

Storage and reagent shelf life
The test paper is stable in ambient conditions, but should not be used around or below 0°C.

Test for lead using potassium iodide and electrolysis

Purpose
To determine the presence of lead in metal objects.

Principle
Small quantities of lead ions are dissolved by electrolysis onto filter paper soaked in nitric acid (HNO$_3$) and then tested with potassium iodide (KI). The filter paper turns bright yellow if lead ions are present. By using pointed paper test strips the area exposed to electrolysis may be greatly reduced. This test is probably the most sensitive of all the lead tests and it has been found to work well in almost all lead testing situations including those of lead-alloyed metals such as bronze or brass with lead content as low as 2% (see p. 30).

$$Pb\ (s)\ +\ HNO_3\ (aq)\ \rightarrow\ Pb^{2+}\ (aq)\ \xrightarrow{KI\ (aq)}\ PbI_2\ (s)$$
$$(yellow)$$

Equipment
- electrolysis cables
- balance (to weigh 0.1g)
- filter paper
- acid-resistant reagent containers
- cotton swabs
- graduated cylinder
- spot-test plate
- 6V battery
- droppers
- microspatula
- stainless steel tweezers

Reagents and safety
- **Acetone:** flammable, and irritant; slight health warning
- **Nitric acid (HNO$_3$):** corrosive, oxidizing, and toxic; severe health rating
- **Potassium iodide (KI):** irritant; moderate health rating
- **Water** [distilled][tap]

Protection
Wear goggles, gloves and protective clothing while handling HNO$_3$.

Other tests to consider or confirm results
- Test for lead using spot-test papers (p. 66).
- Test for lead using potassium dichromate (p. 70).

Reference
Gedye, Ione, Henry Hodges and Andrew Oddy. 1973. *Notes for a Short Course in Conservation.* London: British Museum Research Laboratory.

Reagent preparation

- **0.5M nitric acid (HNO_3) solution (1:30):** add 1mL concentrated HNO_3 to sufficient distilled water to bring the total volume to 32mL. **ALWAYS ADD ACID TO WATER!**
- **10% potassium iodide (KI) solution:** dissolve 1g of KI in 9mL distilled water.

Method of sampling

Tests are performed directly on the surface of the object. The nitric acid will cause a black spot to form on some metals, therefore an inconspicuous area should be selected for testing. The alligator clips can scratch the soft metal.

Procedure

1. Degrease an area of the object with acetone and cotton swabs. This important step should not be overlooked. The reaction will not take place unless there is good contact with the metal.
2. Connect the electrolysis cables, attach one clip to the positive pole of the 6V battery and the other end to the object. If the metal is soft, the alligator clip may scratch the object. To avoid this, filter paper soaked in tap water can be used as a pad for the alligator clip. If this method is used, the filter paper must always be wet to conduct the current.
3. Place a drop of the HNO_3 solution on the spot-test plate.
4. Using stainless steel tweezers, dip a piece of filter paper into the HNO_3 solution.
5. Attach the tweezers to the negative pole of the 6V battery with the other length of wire using the remaining alligator clips. It is important that the tweezers hold the filter paper well within the wet part so that current will be conducted by the tweezers to the filter paper and on through the object. Hold the tip of the filter paper to the object for 5 seconds.
6. Remove the filter paper from the object and add one drop of the KI solution to the filter paper.
7. Rinse the test area with distilled water.

Alternative non-electrolysis procedures

1. Using stainless steel tweezers, dip the filter paper in a drop of the HNO_3 solution.
2. Hold the corner of the paper in contact with the object for 30 seconds or, if possible, lay a small HNO_3-soaked filter paper in contact with the object for 30 seconds. For corrosion products and pigments, sprinkle some of the pigments onto HNO_3-soaked filter paper.
3. Add one drop of the KI solution.
4. Swab the artefact with distilled water to remove the chemicals, then dry.

Observations and interpretation

The formation of a bright yellow color on the tip of the filter paper that was in contact with the object indicates the presence of lead ion (Pb^{2+}) in the object. A stronger acid or longer electrolysis will give a very dark colored reaction with lead. Most other metals will turn the test paper brownish red so an orange or brownish red color is not a positive result. The alternative procedure allows testing of materials that cannot be tested using electrolysis, such as powdered pigments, paint *in situ*, and glazes on ceramics. This is a very sensitive test but it requires longer time.

Storage and reagent shelf life

Nitric acid (HNO_3) is stable when stored in a sealed acid-resistant container in ambient conditions. Potassium iodide (KI) is stable in ambient conditions but on prolonged exposure to air it may become yellow. The KI solution is stable in ambient conditions but it should be stored in a dark bottle.

Test for lead using potassium dichromate

Purpose
To determine the presence of lead in objects.

Principle
Lead alloy objects are soft and mark paper readily. If the object has these characteristics, this test should be carried out. Lead ions precipitate as lead chromate (PbCrO$_4$) in the presence of crystals of potassium dichromate.

$$K_2Cr_2O_7 \ (s) \xrightarrow{acid} H_2CrO_4 \ (aq)$$

$$Pb^{2+} \ (aq) + H_2CrO_4 \ (aq) \rightarrow PbCrO_4 \ (s)$$
$$\text{(yellow)}$$

Equipment

- cotton swabs
- dropper
- magnification
- tweezers

Reagents and safety

- **Acetone**: flammable, irritant; slight health rating
- **Glacial acetic acid**: corrosive, toxic and flammable; moderate health warning
- **Potassium dichromate (K$_2$Cr$_2$O$_7$)**: oxidizer, corrosive, carcinogen and mutagen; extreme health rating. All compounds containing chrome are environmental hazards. These should be handled and disposed of according to the MSDS instructions.
- **Water** [distilled]

Protection
Wear goggles, gloves and protective clothing when handling acetic acid and potassium dichromate. Wash hands after use.

Other tests to consider or confirm results

- Test for lead using spot-test papers (p. 66).
- Test for lead using potassium iodide and electrolysis (p. 68).

Reference
Moss A. A. 1956. *The Identification of Metals. Handbook for Museum Curators*, Part B, *Museum Technique*, Section 8, 2–8. London: Museums Association

Reagent preparation
None.

Method of sampling
The test is performed directly on the surface of the object. The acid may etch or stain the surface so it is important to choose an inconspicuous location on the object for testing and to remove the drop of acid as soon as the observations are made. It is possible to take a sample by rubbing a cotton swab on the surface of the object and then to perform the test on the swab.

Procedure

1. Degrease a small area on the surface of the object with acetone on a swab. This step is important and should not be omitted as sometimes a surface coating or dirt is present which will prevent the reaction from taking place.
2. With tweezers place one or two crystals of potassium dichromate on the area being tested.
3. Cover the crystals with a drop of glacial acetic acid.
4. Let it sit about one minute.
5. Add a drop of distilled water.
6. Swab the artefact with distilled water to remove the chemicals, then dry.

Observations and interpretation
With a small lens look for a precipitate of yellow lead chromate around the crystals; its presence confirms that the object contains lead. The reaction for pure uncorroded lead happens immediately when the drop of distilled water is added. Corroded lead may take longer to react. When the potassium dichromate crystals dissolve in the glacial acetic acid, the drop turns slightly yellow. This can easily be confused with a positive reaction, but the positive reaction for lead occurs when the drop of water is added and the lead chromate forms a precipitate that will settle at the bottom of the drop. The test may not work with lead in alloys such as pewter and bronze.

Storage and reagent shelf life
Potassium dichromate crystals are stable in dry ambient conditions. Glacial acetic acid is stable when stored in a sealed, acid-resistant container in ambient conditions.

Test for mercury salts using diphenylcarbazone

Purpose
To determine the presence of mercury (Hg^{++}) salts such as mercuric chloride used as a pesticide on ethnographic artefacts and natural science specimens.

Principle
Mercury salts react with diphenylcarbazone to form an insoluble blue complex.

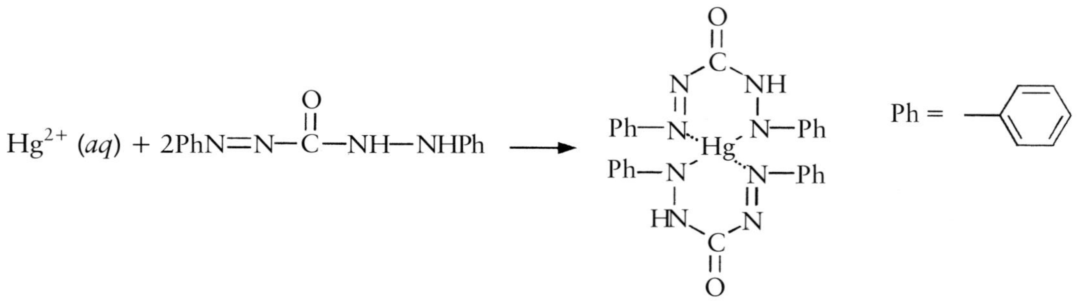

$$Hg^{2+}\,(aq) + 2PhN{=}N{-}\underset{\underset{O}{\|}}{C}{-}NH{-}NHPh \longrightarrow \text{[Hg complex]} \qquad Ph = {-}C_6H_5$$

Equipment

- balance (to weigh 0.1g)
- microspatula
- scalpel
- droppers
- reagent container
- spot-test plate or test tube

Reagents and safety

- **Diphenylcarbazone** [diphenylcarbazide, and diphenylcarbohydrazide will also work]: irritants; slight health rating
- **Ethanol** [ethyl alcohol]: flammable and an irritant; slight health rating
- **Mercury compounds:** highly toxic, irritants and mutagens: extreme (poison) health rating

Protection
None.

Other tests to consider or confirm results

- Test for mercury using aqua regia (p. 74).

Reference
Feigl, Fritz, and Vinzenz Anger. 1972. *Spot Tests in Inorganic Analysis*, 6th English edition. New York: Elsevier.

Reagent preparation

- **Diphenylcarbazone solution**: dissolve 0.1g diphenylcarbazone in 10mL ethanol.

Method of sampling

A small sample must be removed for testing by collecting fallen debris or by wiping the surface with a swab.

Procedure

1. Place a small amount of test material on a spot-test plate or in a test tube.
2. Add 2–3 drops of the diphenylcarbazone solution.

Observations and interpretation

The formation of a blue color indicates the presence of soluble mercuric salts.

Storage and reagent shelf life

Diphenylcarbazone solution should be freshly made before each use.

Test for mercury using aqua regia

Purpose
To determine the presence of mercury in a red pigment or mineral and by inference identify cinnabar (vermilion) (HgS).

Principle
Soluble mercury salts rapidly corrode activated aluminum foil.

$$\text{mercury salts (in pigments)} \xrightarrow[\text{(acid)}]{\text{aqua regia}} Hg^{2+} (aq)$$

$$3Hg^{2+} (aq) + 2Al (s) \rightarrow 3Hg (s) + 2Al^{3+} (aq)$$
$$\text{(foil)} \quad \text{corroded foil will appear dark gray and voluminous}$$

Equipment

- aluminum foil
- emery cloth
- graduated cylinder
- test tube
- balance (to weigh 0.1g)
- filter paper
- microspatula
- blotting paper
- fume hood
- acid-resistant reagent container
- droppers
- heat source (infrared lamp or lab oven)
- spot-test plate or watch glass

Reagents and safety

- **Concentrated hydrochloric acid (HCl):** toxic and corrosive; severe health rating
- **Concentrated nitric acid (HNO_3):** corrosive, oxidizing, and toxic; severe health rating
- **Sodium hydroxide (NaOH) solution:** toxic, corrosive, and an irritant; severe health rating
- **Water** [distilled]
- **Mercury compounds:** highly toxic, irritants and mutagens; extreme (poison) health rating

Protection
Wear gloves, goggles, and protective clothing while handling HCl, NHO_3 and NaOH.

Other tests to consider or confirm results

- Test for mercury salts using diphenylcarbazone (p. 72).
- Test for mercury-cinnabar using a Mel-temp™ apparatus (p. 76).

Reference
Schramm, Hans-Peter. 1995. *Historische Malmaterialien und Ihre Identifizierung.* Stuttgart: Ferdinand Enke Verlag.

Reagent preparation

- **Aqua regia**: start with one part distilled water (5mL is a good amount); add one part (5mL) concentrated nitric acid (HNO_3), then add three parts (15mL) hydrochloric acid (HCl). **ALWAYS ADD ACID TO WATER!** This solution will turn slightly yellow and chlorine gas will be evolved, therefore it should be prepared and used in a fume hood or a well-ventilated area. Place this solution in an acid-resistant storage bottle.
- **10% (w/v) NaOH aqueous solution**: dissolve 1g NaOH in 9mL distilled water. **CAUTION: THIS REACTION IS EXOTHERMIC – THE SOLUTION WILL GIVE OFF HEAT!**

Method of sampling
A small sample is removed to be tested.

Procedure

1. Dissolve a small amount of the sample (a spatula tip full) in a watch glass with a few drops of aqua regia.
2. Evaporate to dryness under a heat lamp or in a lab oven.
3. Dissolve the crystalline residue in a drop of distilled water.
4. Abrade a small spot on a piece of aluminum foil with the emery cloth.
5. Etch the abraded spot by placing a drop of the NaOH solution on it for about a minute. The drop should become milky white because of the evolution of H_2 gas.
6. Dry the foil with blotting paper and carefully dispose of the blotting paper.
7. Deposit one drop of the sample solution on the prepared aluminum surface.

Observations and interpretation
If mercury is present the aluminum will corrode rapidly. The corrosion product is dark gray and voluminous (i.e. looks like a curdy mass). Corrosion of the aluminum is proof of the presence of mercury, and, by inference, of the identity of the sample as cinnabar (vermilion).

Storage and reagent shelf life
Hydrochloric and nitric acids are stable when stored in sealed, acid-resistant containers in ambient conditions. Aqua regia will evolve chlorine gas (Cl_2) during storage. The water component in the above formula for aqua regia allows the mixture to be stored for a somewhat longer period of time, but should not be kept for more than a few days.

Test for mercury-cinnabar using a melting point apparatus

Purpose
To determine the presence of mercury in a red pigment or mineral and by inference identify cinnabar (vermilion).

Principle
Mercuric sulphide, in the form of the mineral cinnabar or the pigment vermilion, will turn black when heated to approx. 290°C (550°F). If sufficient heat (more than 290°C) is applied, metallic mercury will distill and condense out providing that the mineral or pigment contains an appreciable amount of mercury. The small globules of mercury are usually visible under low magnification.

$$\underset{\text{cinnabar}}{Hg^{2+}} + \underset{\text{metals in pigment}}{\text{associate trace}} \xrightarrow{\text{high heat}} \underset{\text{metallic mercury}}{Hg\,(l)} + \underset{\text{trace metals}}{\text{oxidized}}$$

Equipment

- melting point apparatus
- electronic thermometer (thermocouple)
- capillary 'melting point' tubes
- watch glass
- magnification (low power), optional

A melting point apparatus is normally used to determine the melting point of solids.

Reagents and safety

- **Mercury compounds:** highly toxic, an irritant, and a mutagen; extreme (poison) health rating

Protection
Because mercury is volatile, this test might be done in a fume hood. Mercury should be disposed of according to the appropriate environmental guidelines.

Other tests to consider or confirm results

- Test for mercury salts using diphenylcarbazone (p. 72).
- Test for mercury using aqua regia (p. 74).

Reference
Fansett, George R. 1934. *Field Tests for the Common Metals*, 6th edition. Bureau of Mines, Mineral Technology Series No. 36, Bulletin No. 136. Tucson: University of Arizona.

Reagent preparation
None.

Method of sampling
A small sample must be removed for testing.

Procedure
1. Place the sample in a capillary tube.
2. Heat in the melting point apparatus to 290°C.
3. Observe any color change.
4. If necessary, empty the capillary tube into a watch glass and examine with low power magnification. It may be necessary to break the capillary tube to remove the contents.

Observations and interpretation
If the red sample turned black, it is likely to be mercuric sulphide. Look for small silver globules on the walls of the capillary tube. If only a little mercury is formed, it will appear as a gray sublimate or coating composed of minute globules. If the capillary tube is emptied into the watch glass, tiny globules can be observed in the lowest point of the watch glass.

Storage and reagent shelf life
None.

Test for nickel using spot-test papers

Purpose
To determine the presence of nickel in a metal object.

Principle
Some of the sample is dissolved in nitric acid and then tested for nickel using dimethylglyoxime. The formation of a red color confirms the presence of nickel.

$$Ni^{2+} \ (aq) + \text{test paper} \rightarrow \text{red complex}$$

Equipment

- dropper
- tweezers, non-metallic
- scissors
- acid-resistant reagent container

Reagents and safety

- **Nickel test papers** [Macherey-Nagel]; no health rating
- **Nitric acid** (HNO_3): corrosive, oxidizing, and toxic; severe health rating
- **Hydrogen peroxide**; 5% is drugstore grade and 30% is reagent grade
- **Water** [distilled]

Protection
Wear goggles, gloves and protective clothing when handling HNO_3.

Other tests to consider or confirm results

Test for nickel using dimethylglyoxime (p. 80).

Reference
Macherey-Nagel and Co. 1982. *Nickel Test Paper for the Rapid Determination of Nickel in Solutions and in Nickel-containing Alloys.* Product Instructions No. Ni/01/0/2.82. Dueren, Germany: Macherey-Nagel and Co.

Reagent preparation

- **2.5M nitric acid (HNO_3) solution (1:5)**: add one volume of concentrated HNO_3 to five volumes distilled water. **ALWAYS ADD ACID TO WATER!**

Method of sampling

This test is performed directly on the surface of the object. The acid will etch the surface of the metal surface, so the test should be carried out on an inconspicuous area.

Procedure

1. Place a tiny drop of the HNO_3 solution on an inconspicuous spot on the object.
2. Wait for about 1 minute.
3. With non-metallic tweezers hold a piece of the test paper and wick the drop on the object into the paper.
4. Swab the artefact with distilled water to remove the chemicals, then dry.

Observations and interpretation

If nickel (Ni^{2+}) is present there will be a purple red outline where the paper is wet. This is a very sensitive test for the presence of nickel. The instructions state that iron (Fe^{2+}), cobalt ions (Co^{2+}) and cupric ions (Cu^{2+}) may interfere. In the case of iron, the red color should disappear when the test paper is re-acidified or the ferrous ion is converted to ferric ion by the addition of a drop of hydrogen peroxide (H_2O_2). For Co^{2+} and Cu^{2+} the false positive can be eliminated by flushing the test paper in dilute ammonia solution for several minutes after the test solution has been applied. This retains the red from the Ni^{2+} but eliminates the red from the Co^{2+} and Cu^{2+}. The concentration of the acid is important in this test.

Storage and reagent shelf life

Nitric acid (HNO_3) is stable when stored in a sealed acid-resistant container in ambient conditions. The nickel test papers are stable in ambient conditions.

Test for nickel using dimethylglyoxime

Purpose
To determine the presence of nickel in a metal object.

Principle
Some of the sample is dissolved in nitric acid and the presence of nickel ions (Ni^{2+}) is determined using dimethylglyoxime. The formation of a red precipitate confirms the presence of nickel.

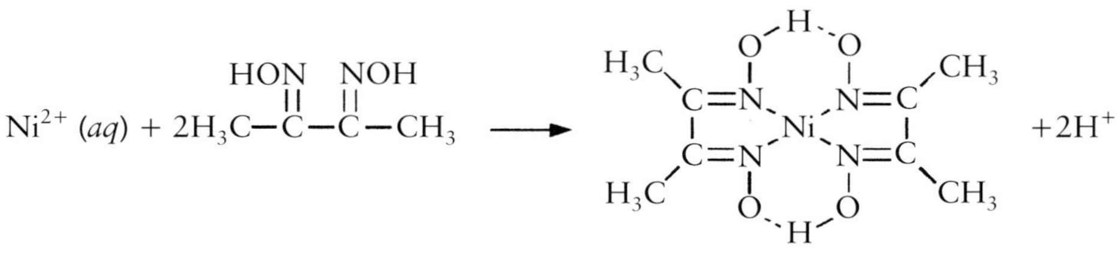

$$Ni^{2+}(aq) + 2H_3C-\underset{\underset{HON}{\|}}{C}-\underset{\underset{NOH}{\|}}{C}-CH_3 \longrightarrow [\text{Ni(dmgH)}_2\text{ complex}] + 2H^+$$

Equipment

- balance (to weigh 0.1g)
- filter paper
- reagent container
- cotton swabs
- graduated cylinder
- spot-test plate
- droppers
- microspatula
- non-metallic tweezers

Reagents and safety

- **Concentrated ammonium hydroxide** (NH_4OH): irritant, and corrosive; severe health rating
- **Dimethylglyoxime** (($CH_3)_2C_2(NOH)_2$): irritant; moderate health rating
- **Ethanol** [ethyl alcohol]: flammable and an irritant; slight health rating
- **Glacial acetic acid** (CH_3COOH): corrosive, toxic, and flammable; moderate health rating
- **Concentrated nitric acid** (HNO_3): corrosive, oxidizing, and toxic; severe health rating
- **Water** [distilled]

Protection
Wear goggles, gloves and protective clothing when handling HNO_3, NH_4OH and CH_3COOH.

Other tests to consider or confirm results
Test for nickel using spot-test papers (p. 78).

Reference
Vogel, Arthur I., and G. Svehla. 1996. *Vogel's Qualitative Inorganic Analysis*, revised. Harlow, England: Longman.

Reagent preparation

- **Dimethylglyoxime solution**: dissolve 0.1g dimethylglyoxime in 10mL ethanol.

Method of sampling

This test is performed directly on the surface of the object. The nitric acid may etch the metal surface leaving a permanent mark, so the test should be carried out on an inconspicuous area.

Procedure

1. Place a tiny drop of concentrated nitric acid (HNO_3) on the surface to be tested.
2. Wait about 10 to 20 seconds.
3. With non-metallic tweezers, hold a piece of the filter paper and wick the drop on the object into it.
4. Clean the area where the acid was with distilled water and dry as soon as possible.
5. Hold the filter paper over an open container of concentrated ammonia (NH_4OH) solution for one minute to neutralize the acid. *Note:* Do not perform this step over the bottle of stock solution as the paper might fall in and contaminate the entire bottle. A small amount may be poured off for use in the test.
6. Add a drop of glacial acetic acid (CH_3COOH) to the filter paper.
7. Add a drop of the dimethylglyoxime solution to the filter paper.

Observations and interpretation

If nickel is present, the filter paper will turn pink or red. This is a very sensitive test for the presence of nickel. If the metal contains a large quantity of copper such as in nickel silver, the copper will turn the filter paper bright blue and the positive for nickel will not be seen. Cobalt ions (Co^{2+}) will interfere to give a brown color. A large excess of iron as ferrous ion (Fe^{2+}) may also interfere.

Storage and reagent shelf life

The dimethylglyoxime is stable for about a year when stored in a sealed container in ambient conditions. Nitric acid and acetic acid are stable when stored in sealed, acid-resistant containers in ambient conditions.

Test for silver using potassium dichromate

Purpose
To determine the presence of silver on the surface of a metal object.

Principle
Sulfuric acid (H_2SO_4) dissolves silver metal to form silver(I) ions. These ions will react with acidified potassium dichromate solution to form silver chromate, which is red and not soluble in water. This test will not distinguish between a silver-plated object and a solid silver object. A second test will also be required to distinguish between a base alloy (>40% copper) or one with a high silver content.

$$2Ag^+ \,(aq) + H_2CrO_4 \,(aq) \rightarrow Ag_2CrO_4 \,(s)$$
$$\text{(red)}$$

Equipment
- droppers
- graduated cylinder
- tweezers
- acid-resistant reagent container
- glass stirring rod

Reagents and safety
- **Potassium dichromate ($K_2Cr_2O_7$)**: oxidizer, corrosive, carcinogens and mutagens; extreme health rating. All compounds containing chromium are environmental hazards. These should be handled and disposed of according to the MSDS instructions
- **Sulfuric acid (H_2SO_4)**: toxic, oxidizing, and corrosive; severe health rating
- **Water** [distilled]

Protection
Wear goggles, gloves and protective clothing when handling H_2SO_4 and potassium dichromate.

Other tests to consider or confirm results
- Test to determine silver quality using nitric acid (p. 86).

Reference
Moss, A. A.1956. *The Identification of Metals. Handbook for Museum Curators*, Part B, *Museum Technique*, Section 8, 2–8. London: Museums Association.

Reagent preparation

- **2M sulfuric acid (H_2SO_4) solution (1:8):** slowly add 3mL concentrated sulfuric acid to 24mL distilled water while stirring continuously. **ALWAYS ADD ACID TO WATER!**

Method of sampling

This test is run directly on the surface of the object. Care should be taken to select an inconspicuous area and the droplet of chemical should be removed, the area swabbed with distilled water as soon as possible and the artefact dried.

Procedure

1. Using tweezers place one or two crystals of potassium dichromate ($K_2Cr_2O_7$) on an inconspicuous area of the object.
2. Place a drop of H_2SO_4 solution on the crystals.
3. Wait for about one minute.
4. Observe the area for the formation of colored crystals and then rinse the area with distilled water and dry.

Observations and interpretation

If a red precipitate of a silver chromate is observed on the surface, silver is present. This test only detects the presence of silver and not the quality of silver. The *Silver quality test* (p. 86) determines if the silver content is high.

Storage and reagent shelf life

Potassium dichromate crystals are stable in ambient conditions. Sulfuric acid is stable when stored in a sealed, acid-resistant container in ambient conditions. Avoid excess heat, and do not store on wood.

Test for silver using potassium dichromate and electrolysis

Purpose
To determine the presence of silver in metal objects.

Principle
Small amounts of silver metal are dissolved by electrolysis onto a piece of filter paper soaked in chromic acid. The chromic acid reacts with silver to form silver chromate (red color). By using pointed paper test strips the area exposed to electrolysis may be greatly reduced (see p. 30).

$$Ag\ (s) \xrightarrow{electrolysis} Ag^+\ (aq)$$

$$2Ag^+\ (aq) + H_2CrO_4\ (aq) \rightarrow Ag_2CrO_4\ (s)$$
$$(red)$$

Equipment

- electrolysis cables
- balance (to weigh 0.1g)
- filter paper
- acid-resistant reagent container
- cotton swabs
- graduated cylinder
- spot-test plate
- 6V battery
- droppers
- microspatula
- stainless steel tweezers

Reagents and safety

- **Acetone**: flammable, irritant; slight health rating
- **Glacial acetic acid**: corrosive, toxic and flammable; moderate health warning
- **Potassium dichromate crystals** ($K_2Cr_2O_7$): oxidizing, corrosive, carcinogens and mutagens; extreme health rating. All compounds containing chromium are environmental hazards. These should be handled and disposed of according to the MSDS instructions
- **Water** [distilled][tap]

Protection
Wear goggles, gloves and protective clothing and wash hands after using potassium dichromate crystals ($K_2Cr_2O_7$).

Other tests to consider or confirm results

- Test to determine silver quality using nitric acid (p. 86).
- Test for silver using potassium dichromate (p. 82).

Reference
Laver, Marilyn. 1978. Spot Tests in Conservation: Metals and Alloys. *International Committee of Museums (ICOM), Committee for Conservation*, 5th Triennial Meeting, Zagreb, 78/23/8:1–11, preprint.

Reagent preparation

- **7% aqueous potassium dichromate solution**: dissolve 1g $K_2Cr_2O_7$ in 14mL distilled water.
- Then add 1mL of glacial acetic acid.

Method of sampling

Tests are performed directly on the object. The alligator clips may scratch or dent soft metals.

Procedure

1. Degrease an inconspicuous area of the object with acetone on a cotton swab. This important step should not be overlooked. The reaction will not take place unless there is good contact with the metal.
2. Connect the electrolysis cables, attach one clip to the positive pole of the 6V battery and the other end to the object. If the metal is soft, the alligator clip may scratch the object. To avoid this, filter paper soaked in tap water can be used as a pad for the alligator clip. If this method is used, the filter paper must always be wet to conduct the current.
3. Place a few drops of the potassium dichromate solution on the spot-test plate. Then, using a pair of stainless steel tweezers, dip a small piece of filter paper into the chromic acid solution. Be sure to hold the filter paper so that one corner or point comes straight out from the tweezers. The filter paper can be properly shaped by cutting it into a triangle with one long side or by folding a square piece of filter paper into a triangle.
4. Attach the tweezers to the negative pole of the 6V battery with the other length of wire using the remaining alligator clips. It is important that the tweezers hold the filter paper well within the wet part so that current will be conducted by the tweezers to the filter paper and on through the object.
5. Hold the corner or point of the filter paper in contact with the object.
6. Remove the filter paper from the object.
7. Rinse the test area with distilled water and then dry.

Observations and interpretation

If silver is present, a small dark red spot will form where the filter paper was in contact with the object, due to the formation of silver chromate, Ag_2CrO_4. Nickel silver (containing no silver) forms a black spot on the surface of the metal.

Storage and reagent shelf life

Potassium dichromate solution is stable in ambient conditions; avoid excess heat, and do not store on wood.

Test to determine silver quality using nitric acid

Purpose
To distinguish between solid silver and silver plating or nickel silver (German silver).

Principle
Nitric acid (HNO_3) dissolves the silver plating to reveal the base metal below. The base metal, usually a copper alloy, will react to form copper nitrates, which are bright green or blue. Nickel silver (sometimes called German silver) contains no silver but has sufficient copper content to react immediately with the nitric acid. Low copper content silver alloys will turn pale cream colored because the copper in the alloy is leached out leaving pure silver on the surface.

$$2Ag\ (s) + 2HNO_3\ (aq) \rightarrow 2AgNO_3\ (aq) + H_2\ (g)$$

(pale green)

Equipment
- droppers
- filter paper
- tweezers
- spot-test plate
- cotton swabs

Reagents and safety
- **Concentrated nitric acid** (HNO_3): corrosive, oxidizing, and toxic; severe health rating
- Water [distilled]

Protection
Wear goggles, gloves and protective clothing while handling HNO_3.

Other tests to consider or confirm results
- Test for silver using potassium dichromate (p. 82).
- Test for silver using potassium dichromate and electrolysis (p. 84).

Reference
Liddicoat, Richard T., Jr., and Lawrence L. Copeland. 1976. *The Jewelers' Manual*. Los Angeles: Gemological Institute of America.

Reagent preparation
None.

Method of sampling
The test is carried out on an inconspicuous area of the object. The acid will etch the surface of the artefact and dissolve the silver plating to reveal the base metal underneath. The acid should be washed off as soon as possible and the area dried. To limit the damage to the surface of the silver artefact caused by the concentrated nitric acid (HNO_3), it is important to limit the area exposed to the acid. For this reason a tiny piece of filter paper is used. The amount of acid delivered to the filter paper must also be controlled so that excess acid does not flow from the filter paper (or from the tweezers) onto the object. The following procedure is recommended to minimize the damage to the artefact.

Procedure
1. Place a drop of concentrated nitric acid (HNO_3) on the spot-test plate.
2. Using tweezers tear off a tiny piece of the filter paper about the size of a pinhead.
3. Using tweezers, carefully dip a corner of the filter paper into the drop of HNO_3.
4. Place the filter paper on an inconspicuous area of the object.
5. After 30–60 seconds observe the color of the filter paper.
6. Wash the acid off the object with distilled water on cotton swabs and then dry.

Observations and interpretation
Silver plate and nickel silver will turn the filter paper green to greenish blue in color immediately. Other silver alloys that are solid silver will turn the filter paper slightly cream colored. The cream color may be more noticeable on the silver metal than the filter paper.

Storage and reagent shelf life
Nitric acid (HNO_3) is stable when stored in a sealed, acid-resistant container in ambient conditions.

Test for tin using cacotheline using electrolysis

Purpose
To determine the presence of tin in a metal object.

Principle
Small amounts of metal are dissolved by electrolysis and then reacted with a solution of cacotheline to form a purple color (see p. 30).

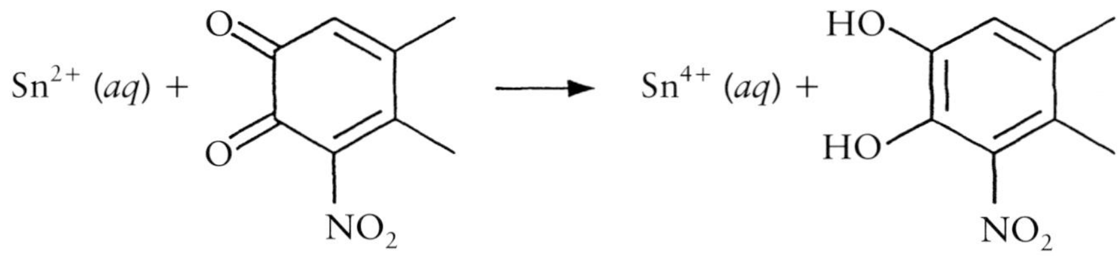

$$Sn^{2+} (aq) + \text{cacotheline fragment} (C_{21}H_{21}N_3O_7 \cdot HNO_3 \cdot H_2O) \longrightarrow Sn^{4+} (aq) + \text{reduced cacotheline fragment (purple)}$$

Equipment

- electrolysis cables
- balance (to weigh 1g)
- filter paper
- acid-resistant reagent container
- cotton swabs
- graduated cylinder
- spot-test plate
- 6V battery
- droppers
- microspatula
- stainless steel tweezers

Reagents and safety

- **Acetone**: flammable, irritant; slight health rating
- **Cacotheline nitrate** ($C_{21}H_{21}N_3O_7 \cdot HNO_3 \cdot H_2O$): irritant; slight health rating
- **Sodium chloride** (NaCl)
- **Hydrochloric acid** (HCl): toxic and corrosive; severe health rating
- **Water** [distilled]

Note: Cacotheline nitrate ($C_{21}H_{21}N_3O_7 \cdot HNO_3 \cdot H_2O$) is a yellow crystalline powder that is a complex organic molecule related to brucine and strychnine.

Protection
Wear gloves, goggles and protective clothing when handling HCl.

Other tests to consider or confirm results
Test for tin using sulfurous acid (p. 90).

Reference
Laver, Marilyn. 1978. Spot Tests in Conservation: Metals and Alloys. *International Committee of Museums (ICOM), Committee for Conservation*, 5th Triennial Meeting, Zagreb, 78/23/8:1–11, preprint.

Reagent preparation

- **Cacotheline solution**: add 1g cacotheline nitrate to 15mL of distilled water. There should always be some crystals of cacotheline at the bottom of the bottle to ensure that the solution is saturated.
- **Saturated sodium chloride (NaCl) solution**: to make a saturated NaCl solution sufficient NaCl should be added to 30mL of distilled water so that crystals of NaCl remain undissolved (it may be necessary to let the container stand overnight).
- **5.8M hydrochloric acid (HCl) solution**: add 7.5mL concentrated HCl to 7.5mL distilled water. ALWAYS ADD ACID TO WATER!

Method of sampling

Tests are performed directly on the object. The cacotheline solution may stain the object so an inconspicuous area should be chosen for testing. The alligator clips can scratch objects, especially soft metals such as tin. Care should be taken when attaching and detaching them.

Procedure

1. Place a drop of the cacotheline nitrate solution on the spot-test plate.
2. Using tweezers, dip a small piece of filter paper into the cacotheline solution, allow most of it to evaporate.
3. Degrease an area of the object with acetone and cotton swabs. This is an important step. The reaction will not take place unless there is good contact with the metal.
4. Connect both ends of the wire to alligator clips, attach one clip to the positive pole of the 6V battery and the other end to the object. If the metal is soft, the alligator clip may scratch the object.
5. Place a drop of the saturated NaCl solution on the filter paper that was previously dipped into the cacotheline solution.
6. Attach the tweezers to the negative pole (cathode) of the 6V battery with the other length of wire using the remaining alligator clips.
7. Using the tweezers hold the tip of the filter paper to the object. It is important that the tweezers are holding the filter paper well within the wet part so that current will be conducted by the tweezers to the filter paper and on through the object.
8. Swab the artefact with distilled water to remove the chemicals, then dry.

Alternative procedure

If the material can be tested with an acid, the following procedure will produce good results.

1. Place a drop of HCl solution on the object for 10 minutes.
2. Wick the drop onto a piece of filter paper.
3. Swab the artefact with distilled water to remove the chemicals, then dry.
4. Place a drop of the cacotheline solution on the filter paper.
5. If there is tin present the paper will turn purple.

Observations and interpretation

The formation of a purple color after 1–2 seconds indicates the presence of tin. In the case of tin alloys (such as pewter) the reactions may take longer (up to 30 seconds).

Storage and reagent shelf life

Hydrochloric acid is stable when stored in a sealed, acid-resistant container in ambient conditions. Cacotheline is stable when stored in ambient conditions. The cacotheline reagent solution remains stable for several months.

Test for tin using sulfurous acid

Purpose
To determine the presence of tin in metal objects.

Principle
Sulfuric acid reacts with sodium hydrosulfite to form sulfurous acid which, in turn, forms sulfur dioxide. In the presence of tin, a reaction occurs with the sulfur dioxide to form black tin sulfide.

$$H_2SO_3\ (aq) \leftrightarrows SO_2\ (g) + H_2O\ (l)$$

$$Sn^{2+}\ (aq) + SO_2\ (g) \rightarrow \underset{\text{(black)}}{SnS\ (s)} + O_2\ (g)$$

Equipment

- cotton swabs
- fume hood
- graduated cylinder
- acid-resistant reagent container
- droppers
- glass stirring rod

Reagents and safety

- **Sodium hydrosulphite** ($NaHSO_3$) [also called sodium bisulphite]: irritant, flammable; slight health rating
- **Concentrated sulfuric acid** (H_2SO_4): toxic, oxidizing, and corrosive; severe health rating
- **Sulfurous acid** (may be used instead of sodium hydrosulphite): toxic, oxidizing and corrosive; severe health rating
- **Water** [distilled]

Protection
Wear goggles, gloves and protective clothing when handling sulfuric and sulfurous acids.

Other tests to consider or confirm results
Test for tin using cacotheline (p. 88).

Reference
Moss, A. A. 1956. *The Identification of Metals. Handbook for Museum Curators*, Part B, *Museum Technique*, Section 8, 2–8. London: Museums Association.

Reagent preparation

- **Solution A** (6M sulfuric acid solution): slowly add 10mL concentrated sulfuric acid (H_2SO_4) to 20mL distilled water while stirring constantly. ALWAYS ADD ACID TO WATER!
  Solution B: dissolve 7.8g of sodium hydrogen sulfite ($NaHSO_3$) in sufficient water to make 100mL of solution. Then add 4mL concentrated sulfuric acid (H_2SO_4) to the solution.
- **Test reagent**: mix 1 part of Solution A with four parts of Solution B. Store in an acid-resistant reagent container.

Note: Sulfurous acid (which is 6% sulfur dioxide in water) may be purchased for use as Solution B. The test reagent solution is prepared by adding 5mL sulfuric acid to 15mL distilled water and then by adding 20mL sulfurous acid.

Method of sampling

The test is performed directly on the surface of the object. The reagent will etch the artefact therefore an inconspicuous area of the object should be selected. The area may need to be polished with a mild polishing cloth after the procedure.

Procedure

1. Place one drop of the test reagent on the surface of the metal and wait 5–8 minutes.
2. Remove the acid from the object with cotton swabs and distilled water and then dry.

Observations and interpretation

If the object is made from tin or a tin alloy (such as pewter), the reagent will form a black ring of tin sulfide around a beige-colored spot.

Storage and reagent shelf life

Solution A is stable when stored in a sealed acid-resistant container in ambient conditions. Solution B ($NaHSO_3$) should not be stored for a long time. The mixture of Solutions A and B should be made fresh each time. If sulfurous acid is purchased for use in this test it should be noted that it gradually oxidizes in air to sulfuric acid.

Test for zinc using ammonium mercuric thiocyanate

Purpose
To determine the presence of zinc in a metal object.

Principle
Zinc reacts to form zinc sulfate ($ZnSO_4$) in the presence of sulfuric acid. Ammonium mercuric thiocyanate reacts with zinc sulfate to form insoluble zinc mercuric thiocyanate, a white precipitate.

$$Zn\ (s) + H_2SO_4\ (aq) \rightarrow ZnSO_4\ (aq) + H_2\ (g)$$

$$ZnSO_4\ (aq) + Hg(NH_4)_2(SCN)\ (aq) \rightarrow ZnHg(SCN)_4\ (s)$$
$$\text{(white)}$$

Equipment

- balance (to weigh 0.1g)
- graduated cylinder
- spot-test plate (dark)
- droppers
- acid-resistant reagent containers
- glass stirring rod

Reagents and safety

- **Ammonium thiocyanate** (NH_4SCN): irritant; moderate health rating
- **Mercuric chloride** ($HgCl_2$): toxic, mutagen/teratogen and irritant; extreme health rating
- **Sulfuric acid** (H_2SO_4): toxic, oxidizing, and corrosive; severe health rating
- **Water** [distilled]

Protection
Wear goggles, gloves and protective clothing when handling H_2SO_4 and mercuric chloride.

Other tests to confirm or consider

- Test for zinc using diphenylthiocarbazone (p. 96).
- Test for zinc in copper alloy using electrolysis (p. 94).

Reference
Moss, A. A.1956. *The Identification of Metals. Handbook for Museum Curators*, Part B, *Museum Technique*, Section 8, 2–8. London: Museums Association.

Reagent preparation

- **3.6M sulfuric acid (H_2SO_4) solution (1:4):** slowly add 4mL of concentrated sulfuric acid to 16mL distilled water while stirring continuously. ALWAYS ADD ACID TO WATER!
- **Ammonium mercuric thiocyanate solution:** dissolve 0.8g mercuric chloride and 0.9g ammonium thiocyanate in 10mL distilled water.

Method of sampling

The test may be carried out directly on the object but, where possible, some material may be removed for testing on the spot-test plate. The acid will etch the artefact so care should be taken to select an inconspicuous area and the droplet of chemical should be removed, the area swabbed with distilled water as soon as possible and the artefact dried.

Procedure

1. Place a small drop of the H_2SO_4 solution on the surface of the object or on the sample.
2. When the evolution of gas bubbles has stopped, transfer the drop of liquid to a dark spot-test plate using a dropper.
3. Swab the artefact with distilled water to remove the chemicals, then dry.
4. With a clean dropper add a small drop of ammonium mercuric thiocyanate solution to the drop on the spot-test plate.

Observations and interpretation

If the sulfuric acid causes vigorous effervescence then the metal is probably zinc. The ammonium mercuric thiocyanate solution confirms the presence of zinc by the formation of a white, feathery precipitate. If no gas evolution is seen then there may be no zinc present or it may be present in such small quantities that it cannot be detected by this test. This test works well for zinc in its pure form – when it is alloyed with other metals it is more difficult to interpret the results.

Storage and reagent shelf life

Ammonium thiocyanate is stable in ambient conditions, it slowly decomposes on exposure to light. Mercuric chloride is stable in ambient conditions but it slowly decomposes to metallic mercury in the presence of organic matter and sunlight. Sulfuric acid is stable when stored in a sealed, acid-resistant container in ambient conditions. The ammonium mercuric thiocyanate reagent solution may be kept for several weeks, but in view of the small amounts needed it may be better to make fresh solutions as required.

Test for zinc in copper alloy using electrolysis

Purpose
To determine the presence of zinc in alloy metallic objects (e.g. brass).

Principle
Small amounts of copper alloy are dissolved by electrolysis and then reacted with ammonium mercuric thiocyanate to form a black or purple compound (see p. 30).

$$Zn\ (s) + Cu\ (s) \xrightarrow{acid} Zn^{2+}\ (aq) + Cu^{2+}\ (aq)$$

$$Zn^{2+}\ (aq) + Cu^{2+}\ (aq) + Hg(NH_4)_2(SCN)_4\ (aq) \rightarrow ZnHg(SCN)_4\ (s) + CuHg(SCN)_4\ (s)$$

(black-purple mixture)

Equipment

- electrolysis cables
- balance (to weigh 0.1g)
- filter paper
- small piece of pure copper
- cotton swabs
- graduated cylinder
- acid-resistant reagent containers
- 6V battery
- droppers
- microspatula
- spot-test plate
- stainless steel tweezers

Reagents and safety

- **Acetone**: flammable, irritant; slight health rating
- **Ammonium thiocyanate** (NH_4SCN): irritant; moderate health rating
- **Mercuric chloride** ($HgCl_2$): toxic, mutagen/teratogen and an irritant; extreme health rating
- **Sulfuric acid** (H_2SO_4): toxic, oxidizing, and corrosive; severe health rating
- **Water** [distilled]

Protection
Wear goggles, gloves and protective clothing when handling mercuric chloride and H_2SO_4.

Other tests to consider or confirm results

- Test for zinc using diphenylthiocarbazone (p. 96).
- Test for zinc using ammonium mercuric thiocyanate (p. 92).

Reference
Moss, A. A. 1956. *The Identification of Metals. Handbook for Museum Curators*, Part B, *Museum Technique*, Section 8, 2–8. London: Museums Association.

Reagent preparation

- **3.6M sulfuric acid (H_2SO_4) solution (1:4):** slowly add 4mL of concentrated sulfuric acid to 16mL distilled water while stirring continuously. ALWAYS ADD ACID TO WATER!
- **Ammonium mercuric thiocyanate solution:** dissolve 0.8g mercuric chloride and 0.9g ammonium thiocyanate in 10mL distilled water.

Method of sampling

Tests are performed directly on the object. The acid will etch the artefact so care should be taken to select an inconspicuous area and the droplet of chemical should be removed, the area swabbed with distilled water as soon as possible and the artefact dried.

Procedure

1. Degrease an area on the object with acetone and swabs. This is an important step. The reaction will not take place unless there is good contact with the metal.
2. Using tweezers, dip a small piece of filter paper into a small drop of the H_2SO_4 solution.
3. Place the filter paper on an inconspicuous area of the artefact.
4. Connect the electrolysis cables, attach one clip to the positive pole of the 6V battery and the other end to the object. If the metal is soft, the alligator clip may scratch the object.
5. Attach a piece of pure copper to the alligator clip that attaches to the negative pole of the 6V battery.
6. Touch the copper to the filter paper for 15 seconds.
7. Remove the filter paper to the spot-test plate.
8. Add a small drop of ammonium mercuric thiocyanate solution to the filter paper.
9. Rinse the area tested with distilled water and then dry.

Observations and interpretation

If zinc is present in the alloy, the filter paper will turn black when the ammonium mercuric thiocyanate solution is added. Pure zinc does not give a black positive with this test. The higher the zinc content, the darker and quicker the black color will appear. During electrolysis the copper ions in the alloy usually turn the filter paper bright yellow. The black appears over the yellow and sometimes looks greenish. If the test is carried out on zinc in association with iron (i.e. galvanized steel), the filter paper will turn dark purple and it will take about 30 seconds for the color to appear. Ignore any dark spots or red or brown colors that appear before the ammonium mercuric thiocyanate solution is added. These are not positives for this reaction.

Storage and reagent shelf life

Ammonium thiocyanate is stable in ambient conditions, it slowly decomposes on exposure to light. Mercuric chloride is stable in ambient conditions, but it slowly decomposes to metallic mercury in the presence of organic matter and sunlight. Sulfuric acid is stable when stored in a sealed, acid-resistant container in ambient conditions. The ammonium mercuric thiocyanate reagent solution can be kept for several weeks, but in view of the small amounts needed, it may be best to prepare a fresh solution for each use.

Test for zinc using spot test papers

Purpose
To determine the presence of zinc in metal, paint, pigment, or pesticides.

Principle
The surface of the object is tested with a specially treated paper that changes color if a significant amount of zinc is present.

$$\text{Zinc} + \text{test paper} \rightarrow \text{red complex}$$

Equipment
- dropper
- scissors
- Erlenmeyer flask
- tweezers
- test tubes

Reagents and safety
- **Quantofix® Zinc Test kit** (includes papers and 32% sodium hydroxide solution) [manufactured by Macherey-Nagel D-5160 Dueren, Germany, distributed in USA by Gallard-Schlesinger Industries, Inc. Garden City, NY 888-686-3454]
- **Quantofix® Zinc Test strips** are an irritant; slight health rating
- **Zink-1** (32% sodium hydroxide, NaOH): corrosive, toxic; severe health rating
- **Water** [de-ionized]

Protection
None.

Other tests to consider or confirm results
- Test for zinc using ammonium mercuric thiocyanate (p. 92).
- Test for zinc in copper alloy using electrolysis (p. 94).

Reference
Macherey-Nagel and Co. n.d. *Quantofix® Zinc Test paper kit*. Product Instructions No. 12738/9.5.7.19/913 10/0681. Dueren, Germany: Macherey-Nagel and Co.

Reagent preparation
None.

Method of sampling
This test can be performed by placing the test paper directly on the surface of the object or by removing sample particles from the object and placing them between a folded piece of test paper. If testing directly on a metal surface, use an inconspicuous area. Alternatively, the object can be sampled by rolling fine cotton swabs dampened with distilled water on several areas of the specimen, especially cracks and crevices.

Procedure

For surfaces and particles

1. Take a test paper and dampen with a small amount of de-ionized water.
2. Add 1 drop of the Zink-1 (32% sodium hydroxide solution) to the test paper.
3. Place the test paper directly on the specimen surface for about 20 seconds OR place the sample particles between a folded piece of test paper.
4. Examine the paper for color reaction immediately.
5. Use a damp swab to clear the test area of reagent residue.

For pesticide residues

1. If swabs were used to gently remove the sample residue from an object surface, the cotton tips should be broken off and placed in an Erlenmeyer flask with 25mL of distilled water. After an hour, place 5mL of this solution into the kit's test tube.
2. Add 10 drops of the Zink-1 (32% sodium hydroxide solution).
3. Place the test paper briefly into the test solution.
4. Wait 30 seconds.
5. Examine the paper for color reaction.

Observations and interpretation
The presence of zinc is indicated if the paper turns pinkish-red quickly. Small quantities of zinc may show color changes unevenly.

Storage and reagent shelf life
Zink-1 reagent (sodium hydroxide) must be stored in a tightly sealed container. The zinc test papers are stable in ambient conditions.

5 Spot tests for inorganic and ionic materials

Calcium	100
Carbonate	102
Chloride/chlorine	104
Nitrate	112
Phosphate	116
Sulfate	122

Test for calcium using nitric acid and sulfuric acid

Purpose
To determine the presence of calcium ions (Ca^{2+}) in a sample. This test can be used in conjunction with the *Test for carbonate using hydrochloric acid and barium hydroxide* (p. 102) to identify calcium carbonates such as those found in lime plaster or in accretions on artefacts.

Principle
Calcium ions (Ca^{2+}) in solution are treated with sulfuric acid and the formation of characteristic calcium sulfate ($CaSO_4 \cdot 2H_2O$)(gypsum) crystals is observed through a hand lens or microscope.

$$Ca^{2+} (aq) + SO_4^{2-} (aq) \rightarrow CaSO_4 \cdot 2H_2O (s)$$
$$\text{(crystals)}$$

Equipment
- glass slide and cover slip
- heat source (lab oven, infrared lamp or hot plate)
- acid-resistant reagent container
- droppers
- graduated cylinder
- magnification (≥20×)
- test tubes

Reagents and safety
- **Nitric acid** (HNO_3): toxic, oxidizing and corrosive; severe health rating
- **Sulfuric acid** (H_2SO_4): toxic, oxidizing and corrosive; severe health rating

Protection
Wear goggles, gloves and protective clothing when handling HNO_3 and H_2SO_4.

Other tests to consider or confirm results
None.

Reference
Schramm, Hans-Peter. 1995. *Historische malmaterialien und ihre identifizierung*. Stuttgart: Ferdinand Enke Verlag.

Reagent preparation

- **0.5M nitric acid (HNO_3) solution (1:30)**: add 1mL concentrated HNO_3 to 30mL distilled water. **ALWAYS ADD ACID TO WATER!**
- **2M sulfuric acid (H_2SO_4) solution (1:8)**: slowly add 5.5mL concentrated sulfuric acid to distilled water while stirring, making up to 50mL of solution. **ALWAYS ADD ACID TO WATER!**

Method of sampling

The sample must be in solution. If solid, place a small amount of the sample into a test tube and add 5–10 drops of the HNO_3 solution to dissolve the calcium-containing part of the sample.

Procedure

1. Place a drop of the sample solution on a glass slide
2. Evaporate to dryness in a lab oven (under the infrared lamp or carefully on a hot plate).
3. Cover the residue with a cover slip.
4. Place a drop of the H_2SO_4 solution next to the cover slip so that it is drawn under the cover slip.
5. Observe the edge of the residue with magnification (loupe or microscope).

Observations and interpretation

As the sulfuric acid dissolves some of the calcium-containing residue, calcium sulfate will precipitate and form characteristic gypsum needles. This occurs usually at the edge of the slide somewhere near the dried residue. It may take as long as 30 minutes for sufficient crystals to form so that they are visible. Not all of the residue will dissolve in the sulfuric acid. The crystals are long and needle-like and may group together in star-like clusters. The crystals are not visible to the naked eye. A microscope with at least 20× magnification works best, however, a magnifying glass/loupe at 15× or 20× may also work. As it is difficult to recognize these crystals the first time, it is advisable to perform this test on a known sample containing calcium.

Storage and reagent shelf life

Sulfuric acid (H_2SO_4) is stable when stored in a sealed, acid-resistant container in ambient conditions. Nitric acid (HNO_3) is stable when stored in a sealed, acid-resistant container in ambient conditions.

Test for carbonate using hydrochloric acid and barium hydroxide

Purpose
To determine the presence of carbonate ions (CO_3^{2-}) such as calcium carbonate ($CaCO_3$) [some plasters, onyx marble, deposits in and on ceramics, soil], basic lead carbonate ($PbCO_3 \cdot Pb(OH)_2$) [a corrosion product found on lead], basic cupric carbonate ($CuCO_3 \cdot Cu(OH)_2$) [a corrosion product found on copper alloys] and other carbonate-containing materials.

Principle
When an acid is added to a material containing carbonate ions (CO_3^{2-}), carbon dioxide gas (CO_2) is evolved.

$$CO_3^{2-} (s) + 2HCl\ (aq) \rightarrow CO_2\ (g) + H_2O\ (l) + 2Cl^-\ (aq)$$

To confirm the presence of carbon dioxide gas (CO_2), a solution of barium hydroxide ($Ba(OH)_2$) is suspended over the effervescing sample. If carbon dioxide gas (CO_2) is present the solution will become cloudy.

$$CO_2\ (g) + Ba(OH)_2\ (aq) \rightarrow BaCO_3\ (s) + H_2O\ (l)$$
$$\text{(cloudy)}$$

Equipment
- balance (to weigh 0.1g)
- glass rod
- magnification
- graduated cylinder
- acid-resistant reagent container
- droppers
- microspatula
- test tube or dark spot-test plate

Reagents and safety
- **Barium hydroxide** ($Ba(OH)_2$): toxic and an irritant; severe health rating
- **Hydrochloric acid** (HCl): toxic and corrosive; severe health rating
- **Water** [boiled, distilled]

Protection
Wear goggles, gloves and protective clothing when using HCl and barium hydroxide.

Other tests to consider or confirm results
None.

Reference
Sorum C. H., 1960. *Introduction to Semimicro Quantitative Analysis*, 3rd edition. Englewood Cliffs, NJ: Prentice-Hall.

Reagent preparation

- **5.8M hydrochloric acid (HCl) solution (1:1):** add 7.5mL concentrated HCl to 7.5mL distilled water. **ALWAYS ADD ACID TO WATER!**
- **3% barium hydroxide (Ba(OH)$_2$) aqueous solution:** add 1.5g Ba(OH)$_2$ to 50mL boiled water. Distilled water should be boiled for at least 5 minutes to remove all of the carbon dioxide (CO$_2$) (see p. 179). This important step should not be ignored. The water should be allowed to cool in a tightly lidded container to prevent more CO$_2$ from dissolving in it while it is cooling. De-ionized water may be used if distilled water is not available but it also must be boiled before use.

Method of sampling

A small sample must be removed for testing. The wash from the surface of an object may be used if it is first evaporated in a watch glass.

Procedure

1. Place a small amount of the sample in a test tube or on a dark spot-test plate.
2. Add two to three drops of the HCl solution.
3. Observe for effervescence.

To confirm the presence of carbon dioxide gas (CO$_2$) within a test tube

Suspend a drop of the barium hydroxide solution on the end of a glass rod or dropper. Insert the drop into the test tube while the contents are still bubbling. If the drop becomes cloudy, the gas is probably CO$_2$; interference can only occur because of the presence of sulfite, which is unlikely in an archaeological setting.

Observations and interpretation

If carbonate ions (CO$_3^{2-}$) are present, the solution will effervesce strongly. It is important to observe the reaction closely, as some materials will dissolve in the acid without effervescing. Just because a material dissolves in the acid does not mean that it contains carbonate ions (CO$_3^{2-}$). Carbon dioxide gas must be evolved for this test to be positive. Some metals such as zinc will also evolve gas when in contact with an acid. In this case, the gas evolved is hydrogen (H$_2$), not carbon dioxide. If the evolution of a gas which is not carbon dioxide is suspected, then the test to confirm the presence of carbon dioxide should be carried out. Note: The drop of Ba(OH)$_2$ may turn cloudy with prolonged contact with air.

Storage and reagent shelf life

Barium hydroxide is stable in ambient temperatures; it will react vigorously with acids. Hydrochloric acid (HCl) is stable when stored in a sealed, acid-resistant container in ambient conditions.

Test for chloride ions using sulfuric acid

Purpose
To establish the presence of chloride ions (Cl⁻) in inorganic compounds such as ceramics, metals, bleached items, or table salt. This test is a good alternative to the silver nitrate test, especially if the sample is in its crystalline (dry) form. This test only works with solid chlorides and concentrated acid, and also detects other anions (e.g. acetate).

Principle
Concentrated sulfuric acid (H_2SO_4) will react with chloride salts to form volatile hydrogen chloride (HCl). The evolution of the latter can be shown by its reaction with pH paper.

$$Cl^- \,(s) + H_2SO_4 \,(l) \rightarrow HSO_4^- \,(aq) + HCl \,(g)$$

$$HCl \,(g) + \text{indicator} \rightarrow [\text{acidified indicator}]^+$$

$$\text{(red to yellow)}$$

Equipment
- Pasteur pipettes
- pH test strips
- microspatula
- small test tube (screw-cap closures work well)
- Parafilm™

Reagents and safety
- **Sulfuric acid** (H_2SO_4): toxic, oxidizing, and corrosive; severe health rating
- **Water** [distilled]

Protection
Wear goggles, gloves and protective clothing when handing sulfuric acid.

Other tests to consider or confirm results
- Test for chloride using silver nitrate (p. 108).
- Test for halogens (chlorine) using pyrolysis (Beilstein test) (p. 106).

Reference
Wiig, Edwin O., Willard R. Line, and John F. Flagg. 1954. *Semimicro Qualitative Analysis*. New York: D. Van Nostrand.

Reagent preparation
None.

Method of sampling
The test is very sensitive so only a small quantity of the sample is required. If the sample is in solution, evaporate about 5mL of the test solution to near dryness.

Procedure

1. Place a few crystals of the material to be tested in a test tube.
2. Add one or two drops (no more!) of concentrated sulfuric acid (H_2SO_4) to the test tube.
3. Quickly place the pH indicator strip in the test tube and seal it. The pH strip must remain in the upper part of the test tube or it will be contaminated by the sulfuric acid. The easiest way to get the pH strip to stay in the upper part of the test tube is to fold it in half, letting it spring open again. The resistance between the two halves of the strip and the wall of the test tube will hold the strip in place. This works well with the indicator strips that are on plastic (ColorpHast brand) as the plastic is very stiff.
4. After about 2 minutes observe the indicator strip for any signs of color change.

Observations and interpretation

If there are chlorides present, the pH indicator strip will indicate more acidic. If there are no chlorides present, the strip will not change at all because sulfuric acid (H_2SO_4) is not volatile. During testing, the pH indicator strips remained unchanged for 24 hours in a test tube with sulfuric acid (H_2SO_4). When testing most chloride-containing salts, the indicator strip changes quickly (within 30 seconds) but for some (barium chloride, stannous chloride) it may take as long as 15 minutes. Corrosion products or impure samples may require a greater quantity of sample to obtain a reaction. Some compounds such as mercuric chloride ($HgCl_2$) do not give a positive result with this test. In some trials, nitrate ions (NO_3^-) gave a false positive result for chlorides, but this was found to occur only after 25 minutes had passed.

Carbonate (CO_3^{2-}) ion-containing materials will cause effervescence as carbon dioxide (CO_2) is evolved, but this will not cause the pH strips to become more acidic unless the sulfuric acid has splashed onto the strip. Ignore any acid changes to the pH paper due to contact with the liquid acid.

Storage and reagent shelf life

Sulfuric acid (H_2SO_4) is stable when stored in a sealed, acid-resistant container in ambient conditions. The pH indicator test strips are stable but should be kept out of direct sunlight.

Test for halogens (chlorine) using pyrolysis (Beilstein test)

Purpose
To determine the presence of chlorine/chlorides, especially for materials being considered for use in long-term conservation or museum applications (e.g. storage materials) such as plastic films, Saran™, or Tygon™ tubing.

Principle
A material containing bound or ionic halogens (chlorine, bromine, iodine) such as salt or poly(vinyl chloride) (PVC), will react with a copper wire when heated in a flame to produce a brilliant, long-lasting green flame.

$$Cu\,(s) + PVC \xrightarrow{heat} CuCl_2 \xrightarrow{flame} \text{green colored flame}$$

Equipment

- single- or multi-strand copper wire (remove several inches of insulation from test end)
- insulated holder (e.g. tongs or rubber grip pliers) is necessary for larger or uninsulated wires
- flame source (alcohol lamp, Bunsen burner)

Reagents
None.

Protection
Open flames in a conservation lab are always a potential hazard. Use extreme caution and wear goggles. There should be no solvent containers in the vicinity or any other materials that can catch fire. Perform the test in a fume hood, if possible.

Other tests to consider or confirm results

- Test for chloride ions using sulfuric acid (p. 104).
- Test for chlorine in polymers using pH paper and pyrolysis (p. 110).

Reference
Williams, R. Scott, 1989. The Beilstein Test: A Simple Test to Screen Organic and Polymeric Materials for the Presence of Chlorine. *CCI Notes* No.17/1. Ottawa: Canadian Conservation Institute.

Reagent preparation
None.

Method of sampling
A small sample of the material must be melted onto a copper wire and burned in a flame (see p. 29).

Procedure

1. Separate out one strand of copper wire for sampling.
2. Clean the wire of contaminants by holding it over the flame for several seconds. It may be necessary to hold the wire with some kind of holder as the entire length of wire will quickly get hot in the flame.
3. While the wire is still hot, dip it into or scrape it over the material to be tested until some of the sample material has coated the wire. If no material melts onto the wire, heat the wire again.
4. Return the wire to the flame and observe the color that appears as the sample material burns.

Observations and interpretation
A strong green color in the flame usually indicates the presence of halogens (chlorine, bromine, iodine but not fluorine). The flame will burn green for a long period of time if chlorinated polymers such as PVC are present. Impurities, such as fingerprints, or surface treatments that contain chloride may give a weak green flame that disappears quickly.

Storage and reagent shelf life
Not applicable.

Test for chloride using silver nitrate

Purpose
To confirm the presence of chloride ions (Cl⁻) in solutions with substances such as accretions on ceramics and corrosion products on metal artefacts. This test may also be used to confirm the presence of chloride ions (Cl⁻) in wash water used to desalinate ceramics.

Principle
Chloride ions (Cl⁻) are dissolved and acidified in a solution of nitric acid (HNO_3), then reacted with silver nitrate ($AgNO_3$) to form silver chloride (AgCl), a white precipitate.

$$Cl^- (aq) + AgNO_3 (aq) \rightarrow AgCl (s) + NO_3^- (aq)$$
$$\text{(white)}$$

Equipment

- balance (to weigh 0.1g)
- graduated cylinder
- scalpel
- cotton swabs
- microspatula
- test tubes or dark spot-test plate
- droppers
- reagent container (dark bottle or cover with foil)
- acid-resistant container

Reagents and safety

- **Nitric acid** (HNO_3): corrosive, oxidizing and toxic; severe health rating
- **Silver nitrate** ($AgNO_3$): toxic, oxidizing and corrosive; severe health rating
- **Water** [distilled]

Protection
Wear gloves, goggles and protective clothing while handling HNO_3 and silver nitrate. Silver nitrate can cause skin and clothing to turn black.

Other tests to consider or confirm results
Test for chloride ions using sulfuric acid (p. 104).

Reference
Sorum C. H. 1960. *Introduction to Semimicro Quantitative Analysis*, 3rd edition. Englewood Cliffs, NJ.: Prentice-Hall.

Reagent preparation

- **0.2M silver nitrate (AgNO$_3$) aqueous solution:** dissolve 1.5g of solid AgNO$_3$ in 50mL of distilled water. Keep in a dark bottle or wrap container with aluminum foil. Prepared silver nitrate solution may also be purchased commercially.
- **7.7M nitric acid (HNO$_3$) solution (1:1):** combine equal quantities of concentrated nitric acid (HNO$_3$) and distilled water. ALWAYS ADD ACID TO WATER!

Method of sampling
A small amount of the sample must be removed for testing.

Procedure

1. Place the sample in a test tube or test plate.
2. Add 5–8 drops of distilled water.
3. Add two drops of the HNO$_3$ solution.
4. Add a drop of the silver nitrate solution.
5. Hold test tube against a black background.

Observations and interpretation
Silver chloride (AgCl) is very insoluble and even small amounts will form a cloudy, white precipitate. Sometimes, only a white haze forms when there are very few chloride ions present. Additional confirmation may be obtained by adding drops of NH$_4$OH solution, which dissolves the precipitate if chloride is present. Interfering ions that also form precipitates, such as iodide, bromide, or sulfide are unlikely to be found in archaeological samples. Large quantities of sulfates can also form a precipitate with silver chloride. Sulfates should be tested for separately.

Note: It is a good idea to run a blank with distilled water for comparison to ensure that the reagents are uncontaminated.

Storage and reagent shelf life
Silver nitrate (AgNO$_3$) is stable when stored in a dark bottle in ambient conditions. Nitric acid (HNO$_3$) is stable when stored in a sealed, acid-resistant container in ambient conditions.

Test for chlorine in polymers using pH paper and pyrolysis

Purpose
To determine the presence of chlorine in materials such as poly(vinyl chloride) (PVC), poly(vinylidene dichloride), or chlorinated rubber.

Principle
Burning chlorine-containing polymers will generate acidic fumes. These fumes react with indicators on pH strips to give an acidic reading.

$$PVC \xrightarrow{combustion} HCl\ (g)$$

$$HCl\ (g) + indicator \rightarrow [H + indicator]^+$$

$$(red\ to\ yellow)$$

$$HCl\ (aq) + AgNO_3\ (aq) \rightarrow AgCl \downarrow + HNO_3$$

Equipment

- flame source (Bunsen burner, alcohol lamp, candle)
- pH paper
- test tube or Pasteur pipette
- test-tube holder
- laboratory wrapping film

Reagents

- **Silver nitrate (AgNO$_3$)**: toxic, oxidizing, severe health rating
- **Water** [distilled]

Protection
Open flames in a conservation lab are always a potential hazard. Extreme caution should be used when using any type of open flame. There should be no solvent containers in the vicinity or any other materials that can catch fire. Wear eye protection when performing this test. The fumes from the burning plastic should not be inhaled. Perform test in well-ventilated area or fume hood if possible. Also note that the plastic is likely to melt onto the bottom of the test tube which renders it useless for further work.

Other tests to consider or confirm results

- Test for halogens (chlorine) using pyrolysis (Beilstein test) (p. 106).

Reference
Braun, Dietrich. 1982. *Simple Methods for Identification of Plastics*. New York: Macmillan.

Reagent preparation

- **0.2M silver nitrate (AgNO$_3$) aqueous solution:** dissolve 0.75g of solid AgNO$_3$ in 25mL of distilled water. Keep in a dark bottle or wrap the container with aluminum foil. Prepared silver nitrate solution may also be purchased commercially.

Method of sampling

A small piece must be removed for testing.

Procedure

1. Place the sample in a test tube (do not place a stopper on the test tube). The Rémillard Pasteur pipette technique may also be used.
2. Place pH indicator strip so that it stays about halfway in the test tube. The easiest way to get the pH strip to stay suspended in the test tube is to fold it in half, letting it spring open again. The resistance between the two halves of the strip and the wall of the test tube will hold the strip in place. This works well with the indicator strips that are on plastic (ColorpHast brand) as the plastic is very stiff.
3. Heat the test tube or Pasteur pipette (use holder) until the sample starts to melt and smoke.
4. Look for a color change in the pH strip that indicates it has become more acidic.
5. If the pH paper shows the presence of acid, add 6 drops of water to the sample and let stand for one minute.
6. Add 2 drops of silver nitrate solution.
7. Look for a cloudy white precipitate, which will indicate the possible presence of chloride.
8. Confirm with NH$_4$OH solution: a few drops will dissolve the white precipitate.

Observations and interpretation

If the sample contains chlorine, the pH strip will become more acidic. This should happen as soon as the smoke from the decomposing plastic reaches the pH strip. The change should be very noticeable going to pH 0 or 1. If there is only a slightly more acidic reading (pH 4 or 5), the sample probably does not contain chloride. In experiments with polyester film, cast nylon film, and polyethylene film, the pH strip became slightly more acidic but not nearly as acidic as the chlorinated plastics tested. However, sulfur and cellulose nitrate will also give an acidic reaction. It is important to do all steps. If the diluted sample contains chlorine, a white precipitate will form with the addition of the silver nitrate solution, and this precipitate will dissolve in NH$_4$OH.

Storage and reagent shelf life

Silver nitrate (AgNO$_3$) is stable when stored in a dark bottle in ambient conditions.

Test for nitrate using spot-test papers

Purpose
To determine the presence of nitrate ions (NO_3) in materials such as soils or organic nitrates in plastics and adhesives (e.g. cellulose nitrate).

Principle
The reaction consists of the reduction of nitrate (NO_3^-) to nitrite (NO_2^-) by a reducing agent. In the presence of acid, the nitrite is converted to nitrous acid (HNO_2) which diazotizes an aromatic amine (sulfanilic acid) and this couples with colorless N[1-naphthyl]ethylenediamine to form a red-violet dye.

$$NO_3^{2-} + \text{reducing agent} \xrightarrow{acid} HNO_2\ (aq)$$

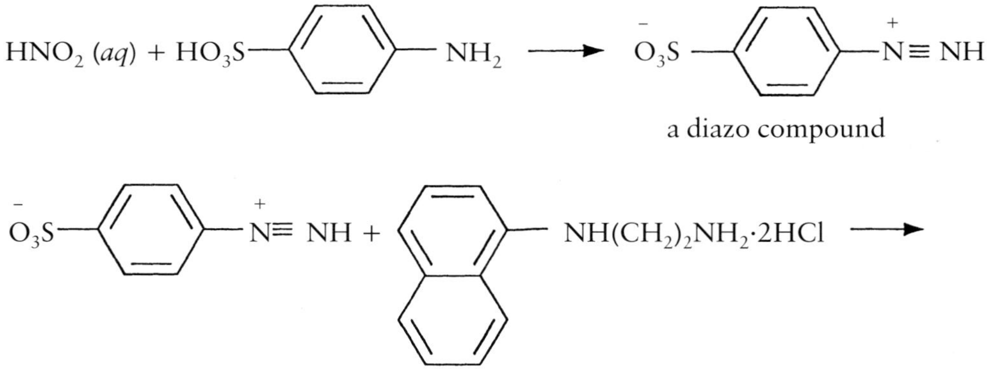

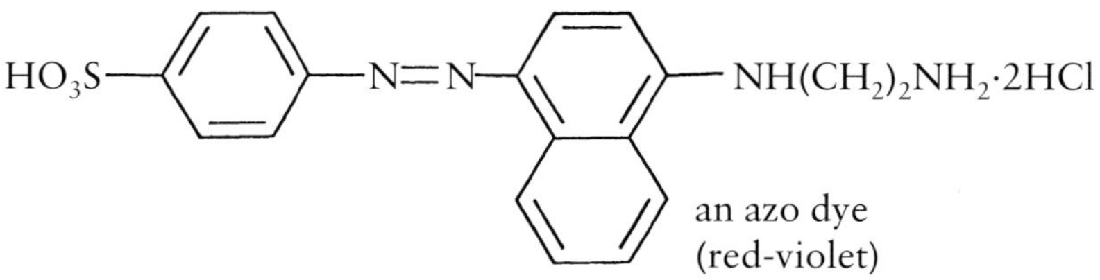

$$HNO_2\ (aq) + HO_3S\text{-}C_6H_4\text{-}NH_2 \longrightarrow {}^-O_3S\text{-}C_6H_4\text{-}N{\equiv}NH^+$$
a diazo compound

$${}^-O_3S\text{-}C_6H_4\text{-}N{\equiv}NH^+ + \text{naphthyl-}NH(CH_2)_2NH_2\cdot 2HCl \longrightarrow$$

$$HO_3S\text{-}C_6H_4\text{-}N{=}N\text{-naphthyl-}NH(CH_2)_2NH_2\cdot 2HCl$$
an azo dye
(red-violet)

Equipment

- droppers or plastic pipettes

Reagents

- **Nitrate (NO_3^-) test strip** (Merckoquant No.1.10020.0001, EM Science); no health rating
- **Water** [distilled or de-ionized]

Protection
None.

Other tests to consider or confirm results

- Test for nitrate (cellulose nitrate) using diphenylamine (p. 164).
- Test for nitrate using iron(II) sulfate (p. 114).

Reference
Merck. nd. Merckoquant Nitrate (NO_3) Test. Product Instructions, Darmstadt, Germany: Merck KgaA.

Reagent preparation
None.

Method of sampling
The test is performed directly on a substrate or may be carried out in dilute aqueous solutions, or with aqueous mixtures involving solids (i.e. soils). Do not touch the paper test strips with the hands. Caution: test papers may stain objects if left in contact for too long a time.

Procedure
Directly on objects
1. Place one or two drops of de-ionized or distilled water on the test strip or water may be place on an inconspicuous area of the object.
2. Hold the test paper strip in contact with the object for 10–30 seconds on the spot where the water was placed. The bulb end on a plastic pipette may help if greater surface contact is needed such as in the case of an irregular or uneven surface.
3. Observe color change in the reaction zone at the end of the test strip.

In aqueous solution
1. Dip test paper strip in solution to be tested for 1 second.
2. Remove excess liquid by shaking and compare reaction color after 1–2 minutes.

In soils
1. Mix a sample of soil with the same quantity of distilled water and filter if necessary.
2. Dip the test paper strip into the clear soil extract and compare the reaction color after 1–2 minutes.

Observations and interpretation
If the reaction zone at the end of the test strip turns purple, nitrate ions are present. Any pink to red-violet coloration indicates the presence of nitrite, which interferes with the reaction (see product instructions to clarify the use of a 10% aqueous amidosulfonic acid solution to eliminate this interference and to clarify problems of pH).

Storage and reagent shelf life
The test strips are stable if kept in a cold (2–8°C or 35–46°F) dry place.

Test for nitrate using iron(II) sulfate

Purpose
To determine the presence of nitrate ions in wash water, soils, or accretions on artefacts.

Principle
The reduction of nitrate to nitric oxide (NO) by iron(II) ions (Fe^{2+}) will form a colored $Fe(NO)^{2+}$ complex.

$$NO_3^- \,(aq) + Fe^{2+} \,(aq) + 4H^+ \,(aq) \rightarrow Fe^{3+} \,(aq) + NO \,(g) + 2H_2O \,(l)$$

$$NO \,(g) + \text{excess } Fe^{2+} \,(aq) \rightarrow Fe(NO)^{2+} \,(aq)$$
$$\text{(brown)}$$

Equipment
- balance (to weigh 1g)
- microspatula
- spot-test plate
- droppers
- acid-resistant reagent container
- test tube
- graduated cylinder
- scalpel

Reagents and safety
- **Sulfuric acid** (H_2SO_4): toxic, oxidizing and corrosive; severe health rating
- **Iron(II) sulfate** (ferrous sulfate, $FeSO_4$): irritant; slight health rating
- **Water** [distilled]

Protection
Wear goggles, gloves and protective clothing when handling H_2SO_4.

Other tests to consider or confirm results
- Test for nitrate (cellulose nitrate) using diphenylamine (p. 164).
- Test for nitrates using spot-test papers (p. 112).

Reference
Hogness, Thorfin R., and Warren C. Johnson. 1954. *Qualitative Analysis and Chemical Equilibrium*, 4th edition. New York: Holt.

Reagent preparation

- **Iron(II) sulfate ($FeSO_4$) aqueous solution**: dissolve 6g $FeSO_4 \cdot 7H_2O$ in 25mL distilled water.

Method of sampling
A small sample must be removed for testing.

Procedure

1. Place a small particle or drop of the sample on the spot-test plate or in a test tube.
2. Place a drop of concentrated sulfuric acid (H_2SO_4) on top of the sample.
3. Place a drop of iron(II) sulfate solution on top of the sample. If a test tube is used it is better to allow the iron(II) sulfate solution to run down the side of the test tube into the sample at the bottom.
4. Look for a brown color forming around the sample.

Observations and interpretation
The presence of a brown ring or cloud around the sample indicates the presence of nitrate or nitrite. Nitrite will turn the solution brown with just the iron(II) sulfate solution. If present, bromide and iodide ions interfere with this test. During trial testing, this test gave a positive for a silver nitrate solution of 1% w/v but not for that of 0.5% w/v. Therefore it may only be sensitive for nitrates in concentrations above 1% w/v.

Storage and reagent shelf life
Sulfuric acid (H_2SO_4) is stable when stored in a sealed, acid-resistant container in ambient conditions. The iron(II) sulfate solution must be freshly made before each use.

Test for phosphate using spot-test papers

Purpose
To determine the presence and concentration range of ortho-phosphate (PO_4^{3-}) ions in a sample.

Principle
The sample is acidified with sulfuric acid and reacted with molybdate ions to form phosphomolybdic acid ($H_7[P(Mo_2O_7)_6]$), which is reduced to phosphomolybdenum blue. The intensity of the color is a semi-quantitative measure of the ortho-phosphate (PO_4^{3-}) concentration. This tests measures only *ortho*-phosphate. All phosphates (*meta*-phosphate (PO_3) and *pyro*-phosphate (P_2O_7)) must be decomposed into *ortho*-phosphate with acid before the total phosphate concentration can be measured accurately. The sample is first acidified with sulfuric acid to decompose all of the phosphates present into *ortho*-phosphate. In this solution, *ortho*-phosphate ions and molybdate ions form phosphomolybdic acid. This is then reduced to form phosphomolybdenum blue. *Note:* a full equation would be too big and complex to show here. The kit reactions are usually proprietary.

$$\text{complex phosphates} \xrightarrow{\text{acid}} PO_4^{3-} (aq)$$

$$PO_4^{3-} (aq) + \text{molybdate ions} \longrightarrow \text{molybdophosporic acid}$$

$$\text{molybdophosporic acid} \xrightarrow{\text{reduction}} \text{phosphomolybdenum blue (blue)}$$

Equipment
- balance (to weigh 0.1g)
- test tubes
- transfer pipettes
- acid-resistant reagent containers

Reagents and safety
- **Hydrochloric acid** (HCl): toxic and corrosive; severe health rating
- **Phosphate Test Kit** (Merckoquant No. 10428, EM Scientific)
- **Sulfuric acid** (H_2SO_4) in kit reagent: toxic, oxidizing and corrosive; severe health rating
- **Water** [distilled]

Protection
Wear goggle, gloves and protective clothing when handling HCl and H_2SO_4.

Other tests to consider or confirm results
- Test for phosphate using ammonium molybdate and benzidine (p. 118).
- Test for phosphate using ammonium molybdate and ascorbic acid (p. 120).

Reference
Persson, K.B. 1997. Soil Phosphate Analysis: A New Technique for Measurement in the Field Using a Test Strip. *Archaeometry* 39(2): 441–3.

Reagent preparation

- **1M hydrochloric acid (HCl) solution (1:11):** add 2mL hydrochloric acid to 22mL distilled water. ALWAYS ADD ACID TO WATER!

Method of sampling
Dissolve or disperse a weighed amount of sample in a known quantity of de-ionized water.

Procedure

1. Clean all glassware and tools with the HCl solution to remove trace quantities of phosphate that could contaminate the test.
2. Place the weighed sample and the known amount of de-ionized water in a test tube.
3. Allow the insoluble material to settle out.
4. Decant the supernatant liquid into a clean test tube.
5. Remove a test strip from container and close immediately.
6. Immerse the reaction zone of the test strip for 1 second in the solution and shake off the excess liquid.
7. Place one drop of the kit reagent labeled (PO_4–1) on the reaction zone of the test strip.
8. Wait for 15 seconds and then shake off excess liquid. CAUTION! THIS REAGENT CONTAINS SULFURIC ACID. CARE MUST BE TAKEN WHEN SHAKING OFF THE EXCESS.
9. Wait for an additional 60 seconds and compare the color on the reaction zone of the test strip with the color chart on the label.

Observations and interpretation
The concentration of the phosphate ions in the solution will be indicated by the color change on the test strip. If the color is more intense than the 500ppm color, dilute the sample and repeat until the color is in the range shown on the label. The test is specific for *ortho*-phosphate ions (PO_4^{3-}) in solution. The only interfering anion is nitrite, which is rare.

Storage and reagent shelf life
Keep the storage tube containing the test strips closed and store between 15° and 25°C. The test kit is labeled with an expiration date.

Test for phosphate using ammonium molybdate and benzidine

CAUTION: Benzidine is a prohibited substance in the UK and can therefore only be used in laboratories specifically licensed by the Health and Safety Executive. For other countries, check with health and safety authorities.

Purpose
To determine if phosphates are present in the sample. The test is very sensitive so even small amounts of phosphate will give a positive reaction. The test may be useful for characterizing soil at an archaeological site (the presence of high concentrations of phosphate in soil might indicate occupation area) or for identifying soluble or insoluble accretions on ceramics.

Principle
Phosphates react with molybdates in mineral acids to form phosphomolybdic acid, which can be observed as a bright yellow precipitate. The precipitate is reacted with benzidine ([1,1-biphenyl]-4,4 diamine) in acetic acid, forming an intense blue color when neutralized with ammonia. The reaction is extremely sensitive as two blue products are formed, benzidine blue and molybdenum blue.

$$PO_4^{3-} (aq) + (NH_4)_6Mo_7O_{24} \cdot 4H_2O \rightarrow (NH_4)_3(PMo_{12}O_{40}) \text{ or } (NH_4)_3[P(Mo_3)_{10})_4]$$

ammonium phosphomolybdate (yellow solid)

$(NH_4)_3[P(Mo_3)_{10})_4]$ + benzidine → benzidine violet (dark blue)

Equipment
- balance (to weigh 0.001g)
- glass stirring rod
- acid-resistant reagent containers
- droppers
- graduated cylinder
- test tube or watch glass
- filter paper
- microspatula
- tweezers

Reagents and safety
- **Ammonium hydroxide** (NH_4OH) [household ammonia will work]: toxic, irritant and corrosive; severe health rating
- **Ammonium molybdate** [$(NH_4)_2MoO_4$]: toxic and an irritant; moderate health rating
- **Benzidine** ([1,1-biphenyl]-4,4 diamine, also known as benzidine free base): toxic, carcinogen; severe health rating
- **Glacial acetic acid** (CH_3COOH): corrosive and flammable; moderate health rating
- **Hydrochloric acid** (HCl): toxic and corrosive; severe health rating
- **Nitric acid** (HNO_3): corrosive, oxidizing and toxic; severe health rating
- **Water** [distilled]

Protection
Wear gloves, goggles and protective clothing when handling NH_4OH, benzidine, HCl and HNO_3.

Other tests to consider or confirm results
- Test for phosphate using spot-test papers (p. 116).
- Test for phosphate using ammonium molybdate and ascorbic acid (p. 120).

Reference
Feigl, Fritz, and Vinzenz Anger. 1972. *Spot Tests in Inorganic Analysis*, 6th English edition. New York: Elsevier.

Reagent preparation

- **8M nitric acid (HNO$_3$) solution (1:1)**: add equal portions of concentrated nitric acid (HNO$_3$) and distilled water. ALWAYS ADD ACID TO WATER!
- **Benzidine solution**: dissolve 0.025g (25mg) benzidine (or its hydrochloride salt) in 5mL glacial acetic acid, and 45mL distilled water. ALWAYS ADD ACID TO WATER!
- **1M hydrochloric acid (HCl) solution (1:10)**: add 5mL concentrated HCl to 50mL distilled water.
- **2M nitric acid (HNO$_3$) solution (1:7)**: add 5mL HNO$_3$ to 35mL distilled water. ALWAYS ADD ACID TO WATER!
- **Molybdate solution**: dissolve 1g ammonium molybdate in 20mL distilled water and add 7mL dilute (8M) HNO$_3$. *Alternative molybdate solution:* as the molybdate solution is only stable for a few days in acid, the following two-part approach is recommended:
- *Part 1*: dissolve 1g of ammonium molybdate in 4mL distilled water, and add 0.8mL concentrated ammonium hydroxide.
- *Part 2*: add 4mL concentrated nitric acid to 6mL distilled water.
  Shortly before use, add four drops of Part 2 to two drops of Part 1. The solution will bubble at first, but when mixed will be transparent yellow.

Method of sampling
A small sample must be removed for testing.

Procedure

1. Place the sample in a test tube or on a watch glass with five drops of the HCl solution.
2. Mix with a glass stirring rod, tap slightly to mix, or wait for ten minutes for any phosphates to dissolve.
3. Deposit one drop of the acid/sample solution on a small piece of filter paper, followed by a drop of the molybdate solution and a drop of the benzidine solution.
4. Hold the filter paper with tweezers over the open mouth of a container of concentrated ammonium hydroxide or household ammonia. Avoid contamination of the stock solution. *Note*: White vapors will emanate from the filter paper as the acid is neutralized by the ammonia vapors.

Observations and interpretation
When the acid on the filter paper has been neutralized, a dark blue spot will form. This should only take a minute or two. The intensity of the blue is dependent upon the phosphate concentration. A faint blue or purple spot may indicate that very little phosphate is present, or that there is contamination of the glassware etc. It is important to note that phosphates are everywhere, in tap water, in soil, on your hands and in many soaps and detergents. Phosphate-free detergents are not necessarily totally PO_4^{3-} free, so some contamination of the sample is likely. Ammonium molybdate will continue to hydrolyze and will form a blue compound. All test results look similar after 5–10 minutes. Also methyl ketones can cause a false positive.

Storage and reagent shelf life
Hydrochloric acid (HCl) is stable when stored in a sealed, acid-resistant container in ambient conditions. Nitric acid (HNO$_3$) is stable when stored in a sealed, acid-resistant container in ambient conditions. The benzidine solutions are stable when stored in sealed containers but must be kept out of the light and apart from oxidizers. The individual parts of the alternative molybdate solutions are also stable when stored in sealed containers at room temperature. The one-part molybdate solution is only stable for a few days.

Test for phosphate using ammonium molybdate and ascorbic acid

Purpose
To determine if phosphates are present in a sample. The test is very sensitive so even small quantities of phosphate will give a positive reaction. The test may be useful for characterizing soil at an archaeological site; the presence of high concentrations of phosphate in soil might indicate an occupation area or soils treated with modern farming fertilizers. This test can also be used to identify soluble or insoluble accretions on ceramics, and to detect phosphate residues from modern detergents.

Principle
Phosphates react with molybdates in mineral acids to form phosphomolybdic acid, which can be observed as a bright yellow precipitate. The precipitate is reacted with ascorbic acid and is reduced to form a blue complex.

$$PO_4^{3-} (aq) + (NH_4)_6Mo_7O_{24} \cdot 4H_2O \rightarrow (NH_4)_3[P(Mo_3O_{10})_4]$$

ammonium phosphomolybdate

$$(NH_4)_3[P(Mo_3O_{10})_4] + \text{Ascorbic acid (vitamins C)} \rightarrow \text{phosphomolybdenum blue (blue)} + \text{L-dehydroascorbic acid}$$

Equipment
- balance (to weigh 0.01g)
- graduated cylinder
- several small ring stands or 100mL beakers
- droppers
- microspatulas
- tweezers
- ≥5cm qualitative filter paper (low ash or no ash)
- acid-resistant reagent containers

Reagents and safety
- **Ammonium molybdate** $((NH_4)_6Mo_7O_{24})$: toxic and an irritant; moderate health rating
- **Ascorbic acid** $(C_6H_8O_6)$ [pure vitamin C]: mild irritant; slight heath rating
- **Hydrochloric acid** (HCl): toxic and corrosive; severe health rating
- **Water** [distilled]

Protection
Wear goggles, gloves and protective clothing when handling HCl.

Other tests to consider or confirm results
- Test for phosphate using spot-test papers (p. 116).
- Test for phosphate using ammonium molybdate and benzidine (p. 118).

Reference
Eidt, Robert C. 1973. A Rapid Chemical Field Test for Archaeological Site Surveying. *American Antiquity* 38(2): 206–11.

Reagent preparation
This test is extremely sensitive, and trace quantities of phosphates are ubiquitous in the everyday world. For this reason all labware used for the preparation and storage of the reagent solutions should first be washed with 1M hydrochloric acid (1:10.6 dilution) and rinsed with de-ionized water.

- **acidified 5% ammonium molybdate solution**: dissolve 2.5g ammonium molybdate in 50mL distilled water and then add 15mL 1:3 dilution hydrochloric acid (HCl).
- **0.5% ascorbic acid solution**: dissolve 0.25g ascorbic acid in 50mL distilled water.

Method of sampling
A small sample must be removed for testing; 0.1g is usually sufficient.

Procedure
1. Place a sheet of filter paper on the ring stand or over the mouth of a beaker so that the bottom of the paper does not touch another surface except at the edges.
2. Using a microspatula, place the sample in the middle of the filter paper. Do not touch the sample with your hands.
3. Place two drops (about 0.1mL) of the ammonium molybdate solution on the sample (the sample should be thoroughly wet), and wait for 30 seconds.
4. Add two drops of the ascorbic acid solution to the sample.
5. Observe results within one minute.

Observations and interpretation
If the sample contains phosphate, a dark blue color will spread outwards from the sample. The rapidity of the reaction and the intensity of the blue are dependent on how extractable the phosphates are and the total phosphate concentration. A faint blue or purple spot may indicate the presence of trace quantities of phosphate, or phosphate-contaminated labware. It is important to note that phosphates are everywhere, in tap water, in soil, on your hands and in many soaps and detergents. A label that indicates 'phosphate-free' detergent does not necessarily mean it is completely free. Apparently, even in the absence of phosphate, the ammonium molybdate will continue to hydrolyze and will form a blue compound. As a result, all test results will look similar after 5–10 minutes, so it is critical to record observations within one minute.

Storage and reagent shelf life
Hydrochloric acid (HCl) is stable when stored in a sealed acid-resistant container in ambient conditions. Ammonium molybdate ((NH_4)$_2MoO_4$) is stable in ambient conditions. The ammonium molybdate solution is stable when stored in a sealed acid-resistant container in ambient conditions. Ascorbic acid ($C_6H_8O_6$) is stable in ambient conditions but the ascorbic acid solution should be made up freshly and discarded if it is 24 hours old.

Test for sulfate using spot-test papers

Purpose
To determine the presence of sulfate ions (SO_4^{2-}) in a solution or on an object; the method yields a semi-quantitative estimate of their concentration.

Principle
The red barium complex of thorin (*o*-[3,6-disulfo-2-hydroxy-1-naphthyl-azo]-benzenearsonic acid) changes to yellow in the presence of an equivalent quantity of sulfate ions. The four squares on the strip contain different quantities of the complex and thus give an estimate of the sulfate concentration.

$$\text{sulfate} + \text{test paper} \rightarrow \text{yellow color}$$

Equipment
- droppers
- tweezers

Reagents
- **Sulfate test papers** (SO_4^{2-}) (Merckoquant, EM Science); no health rating
- **Water** [distilled or de-ionized]

Protection
None.

Other tests to consider or confirm results
Test for sulfate using barium chloride (p. 124).

Reference
Merck KgaA. 1990. Merckoquant Sulfate Test Paper Product Instructions No. 1.10010.001. Darmstadt, Germany: Merck

Reagent preparation
None

Method of sampling
The test may either be performed directly on the surface of an object, or the test strips may be dipped into dilute aqueous solutions, or aqueous mixtures of solids (i.e. soils). Do not touch paper test squares with hands.

Procedure
For object surfaces
1. Place one or two drops of distilled or de-ionized water on the test strip or on an inconspicuous place on the object.
2. Using tweezers, hold the test strip in contact with the object in the area where the water was placed for about 3–5 seconds. Be sure all four test zones on the strip are wet.
3. Compare the resulting color with the chart on the container.

For solutions
1. Dip the test strip in the solution to be tested for 1 second, ensuring that all four test zones are wet. The bulb end on a plastic pipette may help if greater surface contact is needed such as in the case of an irregular or uneven surface.
2. Remove excess liquid by shaking and observe the strip after 1–2 minutes.
3. Compare the resulting color with the chart on the container.

Observations and interpretation
The color chart gives the concentration of sulfate in the test solution. If the test is strongly positive (>1600mg/L), the color change will be bright yellow and immediate. Less positive results become clear in 1–2 minutes. The spot-test papers work best at pH4–8. Refer to the product instructions to clarify problems of pH if they occur.

Storage and reagent shelf life
The test strips are stable if stored below 24°C (75°F) and kept dry.

Test for sulfate using barium chloride

Purpose
To determine the presence of sulfate ions (SO_4^{2-}) in soil, wash water, and accretions on artefacts, etc.

Principle
In an acid solution, sulfate ions (SO_4^{2-}) will form a white, insoluble precipitate with barium ions (Ba^{2+}). The solution must be acidic, otherwise barium salts (other than barium sulfate) will also precipitate giving a false positive.

$$SO_4^{2-}\ (aq) + BaCl_2\ (aq) \rightarrow BaSO_4\ (s) + 2Cl^-\ (aq)$$

(cloudy white)

Equipment
- balance (to weigh 0.1g)
- graduated cylinder
- acid-resistant reagent container
- dark spot-test plate
- droppers
- microspatula
- pH indicator strips

Reagents and safety
- **Barium chloride** ($BaCl_2$): toxic and corrosive; severe health rating
- **Hydrochloric acid** (HCl): toxic and corrosive; severe health rating
- **Water** [distilled]

Protection
Wear goggles, gloves and protective clothing when handling HCl and barium chloride.

Other tests to consider or confirm results
- Test for sulfate using spot-test papers (p. 122).

Reference
Sorum, C. H. 1960. *Introduction to Semimicro Quantitative Analysis*, 3rd edition. Englewood Cliffs, NJ: Prentice-Hall.

Reagent preparation

- **2M barium chloride (BaCl$_2$) aqueous solution 5% (w/v)**: dissolve 1.25g of BaCl$_2$ in sufficient distilled water to make 25mL.
- **3M hydrochloric acid (HCl) solution (1:3)**: add 5mL concentrated HCl to 15mL distilled water. ALWAYS ADD ACID TO WATER!

Method of sampling

A small sample must be removed for testing.

Procedure

1. Place sample on a dark spot-test plate.
2. Add a few drops of distilled water.
3. Add a few drops of the HCl solution.
4. Check with pH indicator paper to make sure the solution is acidic.
5. If it is turbid, allow to settle.
6. Carefully add two drops BaCl$_2$ solution.
7. Wait for a few minutes.
8. Look for a white precipitate or haze.

Observations and interpretation

Barium ions form a white precipitate (clouding) in the presence of sulfate ions. A white haze suggests that some sulfate ions are present; a vigorous precipitate indicates the presence of a lot of sulfate ions. The test is sensitive to 50–70ppm sodium sulfate (Na$_2$SO$_4$). For this test the quantity of the sample matters; if the sample is tiny it may be necessary to reduce the quantity of reagents added.

Storage and reagent shelf life

Barium chloride is stable in ambient conditions. Hydrochloric acid is stable when stored in a sealed, acid-resistant container in ambient conditions. All reagent solutions are stable when stored in sealed containers at room temperature.

6 Spot tests for organic materials

Carbohydrates	128
Fats and oils	136
Organic – animal	140
Organic – vegetable	154
Synthetic polymers	162

Test for starch using iodine/potassium iodide

Purpose
To determine the presence of starch in objects such as paper size, paint, or adhesive.

Principle
A solution of iodine (I_2) in potassium iodide (KI) will turn dark blue if in contact with starch.

$$I_2\ (s) + I^-\ (aq) \rightarrow I_3^-\ (aq)$$

triiodide ion

$$2I_3^-\ (aq) + \text{starch} \rightarrow [\text{starch-}I_5]^- + I^-\ (aq)$$

(blue)

Equipment

- droppers
- balance (to weigh 0.01g)
- graduated cylinder
- reagent container
- capillary tubes
- mortar and pestle
- spot-test plate

Reagents and safety

- **Iodine (I_2):** corrosive, oxidizing and toxic; moderate health rating
- **Potassium iodide (KI):** an irritant; moderate health rating
- **Water** [distilled]

Protection
Wear goggles, gloves and protective clothing when handling iodine.

Other tests to consider or confirm results

- Test for complex carbohydrates using *o*-toluidine (p. 132).
- Test for simple carbohydrates using *o*-toluidine (p. 130).
- Test for carbohydrates using triphenyltetrazolium chloride (p. 134).

Reference
Browning, B. L. 1969. *Analysis of Paper.* New York: Marcel Dekker.

Reagent preparation

- **Iodine/potassium iodide (KI_3) solution:** add 0.9g potassium iodide (KI) to about 5mL distilled water and then add 0.04g of iodine (I_2) to the KI solution. Wait until the iodine is fully dissolved and then dilute to 35mL with distilled water.

Method of sampling

The test can be carried out on an inconspicuous place on the paper or a small piece of the material (a few fibers of paper may be adequate) can be tested. For binders it will be necessary to obtain a small sample of the binder. Pigments in the binder may interfere with the color interpretation.

Procedure

For paper

1. Apply a drop of KI_3 solution to an inconspicuous place on the paper or on a few fibers from the paper.

For binders and other starch samples

1. Grind a small chip of the sample to a fine powder with the mortar and pestle.
2. Place small amount on spot-test plate or in a capillary tube.
3. Place a drop of KI_3 on the sample, or if the capillary tube is used, wick the KI_3 solution into one end.

Observations and interpretation

Formation of a strong blue color indicates the presence of starch; ignore a faint blue color.

Storage and reagent shelf life

The KI_3 solution is stable for about a month if stored in a sealed container away from light. Potassium iodide degrades in air and yellows. Store iodine away from light.

Test for simple carbohydrates using *o*-toluidine

Purpose
To determine the presence of simple carbohydrates (sugars) in objects.

Principle
Dilute sugar solutions react with *o*-toluidine reagent to form a blue-green color. This indicates the presence of simple sugars, such as honey.

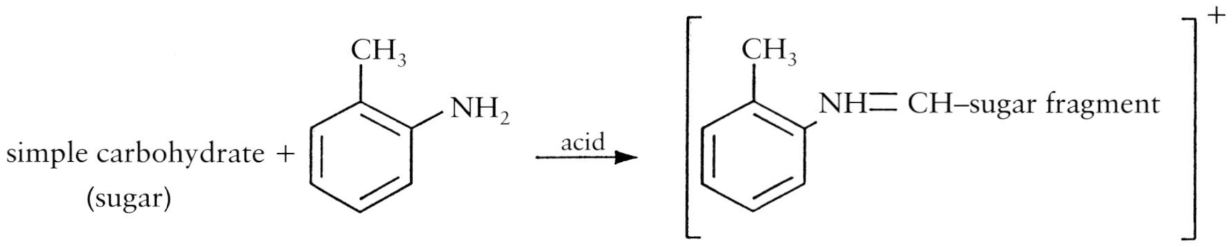

$$\text{simple carbohydrate (sugar)} + \underset{\text{NH}_2}{\underset{\text{CH}_3}{\text{C}_6\text{H}_4}} \xrightarrow{\text{acid}} \left[\underset{\text{NH}=\text{CH--sugar fragment}}{\underset{\text{CH}_3}{\text{C}_6\text{H}_4}} \right]^+$$

(blue-green for hexose or red-brown for pentose)

Equipment

- droppers
- hot plate
- microcentrifuge tubes
- flat-bottomed vials (3mL)
- beaker
- microcentrifuge

Reagents and safety

- *o*-Toluidine reagent from Glucose Testing Kit [Sigma Chemical Co., Cat. No. 365]: toxic, carcinogen, mutagen, and irritant; severe health rating
- **Water** [distilled]

Protection
Wear goggles, gloves and protective clothing when handling *o*-toluidine.

Other tests to consider or confirm results

- Tests for complex carbohydrates using *o*-toluidine (p. 132).
- Test for carbohydrates using triphenyltetrazolium chloride (p. 134).
- Test for starch using iodine/potassium iodide (p. 128).

Reference
Stulik, Dusan, and Henry Florsheim. 1992. Binding Media Identification in Painted Ethnographic Objects. *Journal of the American Institute for Conservation* 31(3): 275–88.

Reagent preparation
None required (the reagents can be used straight from the diagnostic kit).

Method of sampling
If the material to be tested is on an object, a small sample will have to be removed for identification. Use about 5mg of finely ground paint. If there is little solid matter (i.e. a coating or adhesive) then the centrifugal procedure is not necessary.

Procedure

1. Place 5mg of finely ground sample in a 3mL flat-bottomed vial.
2. Add ten drops (about 0.5mL) of distilled water.
3. Boil gently in a boiling water bath on the hot plate for 1–2 minutes.
4. Cool to room temperature.
5. Transfer the liquid to a microcentrifuge tube and spin for about 1 minute.
6. Mix three drops of supernatant with 0.5mL of the *o*-toluidine reagent.
7. Heat in the boiling water bath for 10 minutes or heat for 30 seconds in a microwave oven.

Observations and interpretation
The formation of a blue-green color indicates the presence of simple hexose sugars such as glucose. A red-brown color indicates pentose sugars such as xylose. A very faint color indicates the presence of only traces of sugars, and may be produced by other media, such as plant juices, milk or blood. Charcoal pigment may also interfere, and it is desirable to use non-black paint samples. If the whole kit is purchased, it comes with some glucose standards. These should be used as a control to make sure that the test is working. The reaction does take the full 10 minutes to develop. It starts out as green and becomes bluer the longer it is in the bath. It is also important to note that this test was developed to test for very small quantities of glucose in the blood. With large quantities of sugar present, the color change may be more greenish brown than blue. This is useful for testing small quantities of binding media. If the test is being used to identify accretions or other materials where a large sample can be tested, the concentration should not exceed 1 or 2 percent by weight. Carrying out this test with a series of known samples is very useful in the interpretation of the results. After several hours the color becomes brown.

Storage and reagent shelf life
The *o*-toluidine reagent is labeled with an expiration date. It is stable in ambient conditions; it reacts violently with acids and becomes reddish brown on exposure to air and light. The glucose standards that come with the kit should be refrigerated.

Test for complex carbohydrates using *o*-toluidine

Purpose
To determine the presence of carbohydrates in media, paint and adhesives or deposits. This test identifies complex carbohydrates such as plant gums, starches, or other polysaccharides.

Principle
Complex carbohydrates are composed of simple carbohydrates linked together. If the test for simple carbohydrates is negative or only faintly positive, the sample may be composed of complex carbohydrates, which must first be broken down (hydrolyzed) into simple carbohydrates by the addition of dilute sulfuric acid. The residue is then tested for the presence of simple carbohydrates with *o*-toluidine.

$$\text{complex carbohydrate} \xrightarrow{\text{acid}} \text{simple carbohydrate}$$

$$\text{simple carbohydrate (sugar)} + \underset{}{\text{o-toluidine}} \xrightarrow{\text{acid}} \left[\text{NH=CH-sugar fragment} \right]^{+}$$

(blue-green for hexose or red-brown for pentose)

Equipment

- beaker
- lab oven
- microspatula
- screw-cap vial (3mL)
- droppers
- microcentrifuge
- pH indicator papers
- hot plate
- microcentrifuge tubes
- acid-resistant reagent containers

Reagents and safety

- **Ammonium hydroxide** (NH_4OH, 28–30%): toxic, an irritant, and corrosive; severe health rating
- ***o*-Toluidine from Glucose Testing kit** [Sigma, Cat. No. 365]: toxic, carcinogen, mutagen, and irritant; extreme health rating
- **Sulfuric acid** (H_2SO_4): toxic, oxidizing, and corrosive; severe health rating
- **Water** [distilled]

Protection
Wear gloves, goggles, and protective clothing when handling H_2SO_4, *o*-toluidine, NH_4OH.

Other tests to consider or confirm results

- Test for simple carbohydrates using *o*-toluidine (p. 130).
- Test for carbohydrates using triphenyltetrazolium chloride (p. 134).
- Test for starch using iodine/potassium iodide (p. 128).

Reference
Stulik, Dusan, and Henry Florsheim. 1992. Binding Media Identification in Painted Ethnographic Objects. *Journal of the American Institute for Conservation* 31(3): 275–88.

Reagent preparation

- **0.5M sulfuric acid (H_2SO_4) solution (1:34):** add 1mL concentrated H_2SO_4 to 34mL distilled water. **ALWAYS ADD ACID TO WATER!**
- **7.5M ammonium hydroxide (NH_4OH) solution (1:1):** add equal amounts of concentrated ammonium solution to water.

Method of sampling

If the material to be tested is on an object, a small sample will have to be removed for identification. Use a tiny amount (about 5mg) of finely ground sample. If there is little solid matter (i.e. a coating or adhesive) then the centrifuge procedure is not necessary.

Procedure

1. Place about 5mg sample in a 3mL screw-cap vial.
2. Add 0.5mL (10 drops) of the H_2SO_4 solution and close vial tightly.
3. Heat in a 100°C lab oven for 2 hours.
4. Remove vial from oven, cool and transfer contents to a micro-centrifuge tube.
5. Centrifuge for 1 minute.
6. Place two drops of the supernatant into a screw-cap vial.
7. Add one drop of the NH_4OH solution to neutralize the acid. Test the pH of the supernatant liquid before proceeding to the next step. If it is not neutral or slightly basic, add another drop of NH_4OH solution.
8. Add 0.5mL (10 drops) *o*-toluidine reagent.
9. Place in boiling water bath for 10 minutes or heat for 30 seconds in a microwave oven.

Observations and interpretation

The formation of a blue-green color indicates the presence of simple hexose sugars. If a positive was found only after acid hydrolysis, this indicates that complex carbohydrates are present. Certain pigments [battery black (impure manganese dioxide), red and yellow ochre] have been found to give false negatives. Therefore it best not to use samples containing these pigments. Samples containing gums (arabic, tragacanth) were found to produce a more brownish than blue-green color indicating the presence of pentose sugars. If this test is carried out using a series of known samples, this helps in the interpretation of the results. After several hours the color becomes brown.

Storage and reagent shelf life

Ammonium hydroxide is stable when stored in ambient conditions (avoid heat and sunlight). The *o*-toluidine reagent has an expiration date on the label and is stable in ambient ordinary conditions; it reacts violently with acids and becomes reddish brown on exposure to air and light. Sulfuric acid is stable when stored in a sealed, acid-resistant container away from heat and sunlight.

Test for carbohydrates using triphenyltetrazolium chloride

Purpose
To determine the presence of carbohydrates in media, paint, and adhesives or deposits.

Principle
When polysaccharides containing reducing sugars are hydrolyzed, these may be oxidized by triphenyltetrazolium chloride to form a red complex (Red Formazan).

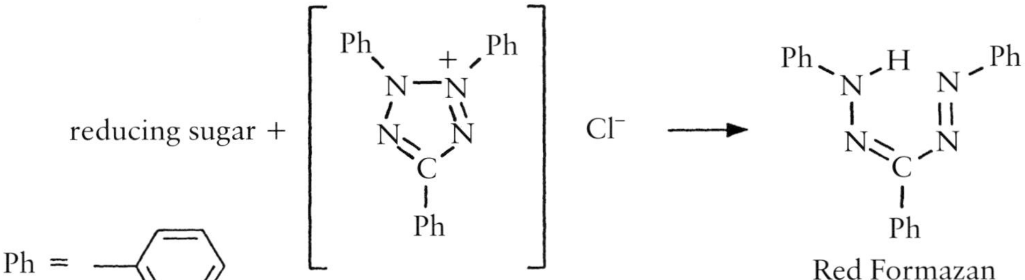

$$\text{reducing sugar} + \begin{bmatrix} \text{Ph}\diagdown \underset{N}{N} \overset{+}{-} \underset{N}{N} \diagup \text{Ph} \\ \text{N} \diagdown \underset{C}{} \diagup \text{N} \\ | \\ \text{Ph} \end{bmatrix} Cl^{-} \longrightarrow \underset{\substack{\text{Red Formazan} \\ \text{(red)}}}{\text{Ph}\diagdown N \overset{H}{-} N \diagup \text{Ph} \atop N \diagdown \underset{C}{} \diagup N \atop | \atop \text{Ph}}$$

$$\text{Ph} = -\bigcirc$$

Equipment

- balance (to weigh 0.1g)
- hot plate
- reagent containers
- droppers
- graduated cylinder
- spot-test plate
- heat source (lab oven or infrared lamp)
- microspatula
- test tube

Reagents and safety

- **Concentrated hydrochloric acid (HCl)**: toxic, corrosive; severe health rating
- **Sodium hydroxide (NaOH)**: toxic, corrosive, irritant; severe health rating
- **Triphenyltetrazolium chloride solution** [TPTZ or TTC]: irritant, moderate health rating
- **Water** [distilled]

Protection
Wear gloves, goggles, and protective clothing while using NaOH and HCl.

Other tests to consider or confirm results

- Test for complex carbohydrates using *o*-toluidine (p. 132).
- Test for simple carbohydrates using *o*-toluidine (p. 130).

Reference
Borelli, E. 1993. *Mural Paintings: Conservation Course-Identification of Binding Media-Lab Notes*, Part I, *Constituent Materials/Execution Techniques*. Rome: ICCROM.

Reagent preparation

- **10% Triphenyltetrazolium chloride solution (TPTZ) solution**: add 1g TPTZ to 9mL distilled water.
- **1.2M NaOH solution**: add 1.25g NaOH to 24mL distilled water.

Method of sampling
A small amount of the material must be removed for testing.

Procedure

1. Place a small amount of the sample on the spot-test plate.
2. Add one drop of concentrated HCl.
3. Evaporate to dryness with the IR lamp or in a lab oven (avoid burning the sample). Add two drops of distilled water to dissolve the hydrolyzed material.
4. Transfer to a test tube with a clean dropper.
5. Add two drops of the TPTZ solution.
6. Add one drop of the NaOH solution.
7. Bring to gentle boil in a water bath for one or two minutes

Observations and interpretation
The formation of a red color or precipitate is evidence of the presence of reducing sugar in the sample.

Storage and reagent shelf life
Hydrochloric acid (HCl) is stable when stored in a sealed, acid-resistant container in ambient conditions. Sodium hydroxide is stable when stored in a sealed container in ambient conditions. The sodium hydroxide solution will absorb carbon dioxide from the air and become less basic if the container is not sealed tightly. Triphenyltetrazolium chloride is stable when stored in a sealed container in ambient conditions.

Test for triglycerides using triglyceride reagent

Purpose
To determine presence of triglycerides in materials such as paint binder, accretions or deposits.

Principle
Triglycerides are esters of carboxylic acids and glycerine. The reagent employed separates the glycerine and, through a series of steps, converts it into a characteristically colored species.

$$\text{fats and oils} + \text{lipoprotein lipase (enzyme)} \longrightarrow \text{glycerol} + \text{fatty acids}$$

$$\text{glycerol} + \text{ATP} \xrightarrow{\text{GK}} \text{glycerol-1-phosphate} + \text{ADP}$$

$$\text{glycerol-1-phosphate} + O_2(g) \xrightarrow{\text{GPO}} \text{dihydroxyacetone phosphate} + H_2O_2$$

$$H_2O_2 + \text{4-aminoantipyrine} + N\text{-ethyl-}N\text{-[3-sulfopropyl]-}m\text{-anisidine sodium salt} \longrightarrow \text{a quinoneimine dye (pink-purple)}$$

Equipment

- syringe
- filter paper

Reagents and safety

- **Triglyceride reagent** (Sigma Diagnostic GPO-Trinder reagent kit, Cat. No. 339): irritant, toxic and possible mutagen; severe health rating
- **Water** [distilled]

Protection
Wear goggles, gloves and protective clothing when handling GPO-Trinder reagent.

Other tests to consider or confirm results

- Test for unsaturated oils using potassium permanganate (p. 138).

Reference
Stulik, Dusan, and Henry Florsheim. 1992. Binding Media Identification in Painted Ethnographic Objects. *Journal of the American Institute for Conservation* 31(3): 275–88.

Reagent preparation
See GPO-Trinder kit instructions.

Method of sampling
A small sample must be removed for testing.

Procedure

1. Obtain a small sample.
2. Place sample (enough to see) onto a narrow rectangular strip of filter paper (like a pH test strip).
3. Use a syringe to extract 1 or 2 drops of GPO-Trinder reagent.
4. Add 1 or 2 drops of GPO-Trinder reagent to the sample. Allow the reagent to run past the sample like a chromatographic process.
5. Observe the color reaction.

For an alternative procedure see Stulik and Florsheim (1992).

Observations and interpretations
The formation of a pink to purple color indicates that the sample contains oils or fats. The GPO-Trinder reagent will stain the paper slightly. A positive reaction occurs when the colored reaction is emanating from the sample itself. This test is best seen with a blank or known negative (distilled water may be used as a known negative).

Storage and reagent shelf life
The GPO-T kit is stable when stored in a sealed container but should be refrigerated because the test contains enzymes.

Test for unsaturated oils using potassium permanganate

Purpose
To determine the presence of olefinic compounds such as unsaturated oils in media, adhesives or accretions.

Principle
Potassium permanganate ($KMnO_4$) oxidizes olefinic double bonds while decolorizing the permanganate and forming manganese dioxide (MnO_2), a brown precipitate.

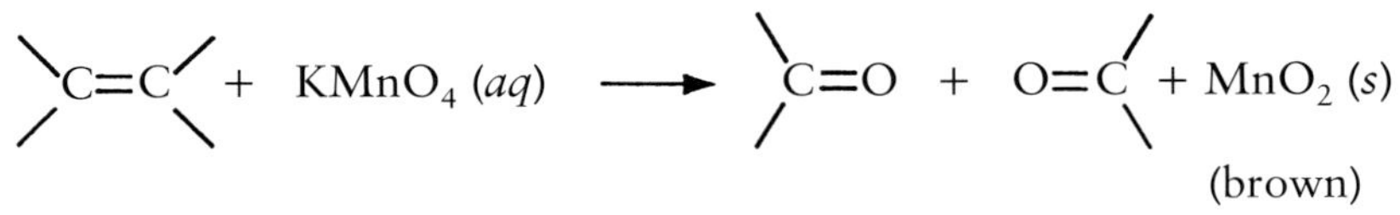

$$\text{>C=C<} + KMnO_4\ (aq) \longrightarrow \text{>C=O} + \text{O=C<} + MnO_2\ (s)$$
$$\text{(brown)}$$

Equipment

- test tube
- droppers
- balance (to weigh 0.1g)
- graduated cylinder
- microspatula
- reagent container

Reagents and safety

- **Potassium permanganate** ($KMnO_4$): oxidizing and corrosive; moderate health rating
- **Ethanol** (ethyl alcohol): flammable and an irritant; slight health rating
- **Water** [distilled]

Protection
Wear goggles, gloves and protective clothing when handling reagents.

Other tests to consider or confirm results

- Test for triglycerides using triglyceride reagent (p. 136).

Reference
Pavia, Donald L., Gary M. Lampman, and George S. Kris. 1982. *Introduction to Organic Laboratory Techniques: A Contemporary Approach*, 2nd edition. Saunders Golden Sunburst Series. Philadelphia: Saunders College Publications.

Reagent preparation

- **1% aqueous potassium permanganate ($KMnO_4$) solution:** dissolve 0.25g $KMnO_4$ in 25mL distilled water.

Method of sampling
Use 25–50mg of substance (two drops if a liquid).

Procedure

1. Place a few drops of the sample in a test tube. If the sample is a solid, dissolve some of it in ethanol.
2. Slowly add the $KMnO_4$ solution one drop at a time while carefully shaking the test tube.

Observations and interpretation
If an unsaturated compound is present, the purple color of the test reagent will change to brown and a brown precipitate will start to form, usually within a minute. Some other readily oxidized compounds, such as aldehydes will also give a positive test. If there is no immediate reaction then the sample is not likely to be an oil containing unsaturated double bonds. After a few minutes, the $KMnO_4$ solution will turn brown on exposure to air.

Storage and reagent shelf life
The potassium permanganate ($KMnO_4$) solution is stable when stored in a sealed container in ambient conditions.

Test for blood using benzidine

CAUTION: Benzidine is a prohibited substance in the UK and can therefore only be used in laboratories specifically licensed by the Health and Safety Executive. In other countries, check with the health and safety authorities.

Purpose
To detect on objects and architecture the presence of blood as a binder, deposit or paint.

Principle
Enzymatic reaction of hydrogen peroxide with blood causes benzidine to change color.

$$H_2O_2\ (aq) + \text{heme} \rightarrow 2[OH]^\bullet$$
$$(C_{23}H_{32}N_4O_4Fe)$$

$$2[OH]^\bullet + H_2N\text{-benzidine-}NH_2 \longrightarrow HN\text{=benzidine violet=}NH + 2H_2O\ (l)$$

benzidine → benzidine violet (dark blue)

Equipment

- droppers
- graduated cylinder
- acid-resistant reagent container
- filter paper
- microspatula
- swabs

Reagents and safety

- **Benzidine** (*p*-Diaminodiphenyl, $NH_2(C_6H_4)_2NH_2$): toxic and a carcinogen; severe health rating
- **Glacial acetic acid** (CH_3COOH): corrosive, toxic and flammable; moderate health rating
- **Hydrogen peroxide** (H_2O_2): irritant; slight health rating

Protection
Wear goggles, gloves and protective clothing when handling benzidine.

Other tests to consider or confirm results
None.

Reference
Bauer, Wilhelm P. 1983. *Class notes. Scientific Principles of Conservation*, Rome: ICCROM.

Reagent preparation

- **Benzidine solution**: add a small amount of benzidine (a microspatula tip full) to 3mL of glacial acetic acid, then add five drops of dilute hydrogen peroxide (H_2O_2) (drugstore variety is 3% and may be substituted). Stronger H_2O_2 solutions will increase reaction but will shorten the shelf life.

Method of sampling

A small sample must be taken or scrapings from the surface may be used. This test is very sensitive so samples may also be taken by transfer to a damp swab or piece of blotting paper for testing.

Procedure

1. Place the sample on a piece of white filter paper, or directly on a swab.
2. Place a drop of benzidine solution on the sample.
3. Observe closely for color to develop around the sample.

Observations and interpretation

A change of color to bluish-green indicates the presence of blood. A sample of fresh blood will cause the reaction to take place immediately and form a dark blue or bluish-green; with a sample of old blood, especially from an archeological sample, the reaction may take longer (up to a minute) and the color may tend to be more green or greenish-yellow. The color fades after a few minutes.

Storage and reagent shelf life

Acetic acid is stable when stored in an acid-resistant reagent container in ambient conditions, but heat, sunlight and freezing may contribute to its instability: store above 17°C (63°F). Benzidine is sensitive to light and should be stored in the dark in a closed amber container. Benzidine solution should be made freshly for each use. Hydrogen peroxide must be reasonably fresh as it is sensitive to light and heat (it can be stored in the refrigerator but it is stable in ambient conditions).

Test for protein (nitrogen) using calcium oxide and pyrolysis

Purpose
To determine the presence of ammonia or amide nitrogen (e.g. protein) in certain materials such as adhesives, deposits or painting media etc.

Principle
Proteinaceous materials contain bound nitrogen which will form ammonia (NH_3) when heated with calcium oxide (CaO). The ammonia formed causes the pH paper to change color indicating the presence of base.

$$\text{protein} + \text{CaO} \,(s) \xrightarrow{\text{heat}} NH_3 \,(g) + \text{residue products}$$

$$NH_3 \,(g) + \text{indicator} \rightarrow [\text{indicator-H}]^-$$

(green to violet)

Equipment

- test tube
- test-tube holder
- pH indicator paper
- source of flame (alcohol burner, Bunsen burner)
- microspatula

Reagents and safety

- **Calcium oxide (CaO)**: an irritant; slight health rating

Protection
Open flames in a laboratory are always a potential hazard. Use extreme caution and wear goggles. There should be no solvent containers in the vicinity or any other materials that can catch fire.

Other tests to consider or confirm results

- Test for protein using copper(II) sulfate (Biuret test) (p. 144).

Reference
Borelli, E. 1993 *Mural Paintings: Conservation Course-Identification of Binding Media-Lab Notes*, Part 1, *Constituent Materials/Execution Techniques*. Rome: ICCROM.

Reagent preparation
None.

Method of sampling
A small sample must be removed for testing

Procedure

1. Place a small amount of sample in a test tube. (This test can also be performed by the Rémillard pipette technique.)
2. Add a microspatula tip full of calcium oxide.
3. Place a pH indicator strip in the mouth of the test tube. The easiest way to make the pH strip stay in the upper part of the test tube is to fold it in half, letting it spring open again. The tension between the two halves of the strip and the wall of the test tube will hold the strip in place. This works well with indicator strips that are supported by plastic (ColorpHast brand), as plastic is very stiff.
4. Hold the test tube with a test-tube holder and heat over the flame.
5. Observe the pH indicator paper.

Observations and interpretation
If the pH paper becomes a more basic (higher pH) color this indicates that ammonia or an amine is present.

Storage and reagent shelf life
The calcium oxide will absorb moisture and carbon dioxide from the air therefore it must be kept in a tightly sealed container.

Test for protein using copper(II) sulfate (Biuret test)

Purpose
To determine the presence of protein in materials such as leather, sinew, vellum, horn, tortoiseshell etc. The test will also identify protein on paper that has been sized with gelatin, paint binders or adhesive, and deposits.

Principle
In the presence of protein the reagents form a purple complex.

$$\text{protein} + CuSO_4\ (aq) \xrightarrow{\text{base}} \text{[Cu-protein complex]}$$

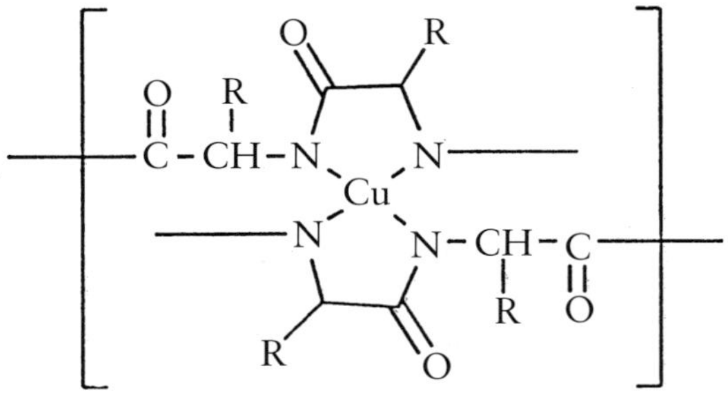

(purple)
R is a carbon containing chain

Equipment

- droppers
- spot-test plate
- blotting paper
- balance (to weigh 0.1g)
- graduated cylinder
- reagent containers
- microspatula
- magnification (optional)

Reagents and safety

- **Copper(II) sulfate** (cupric sulfate, $CuSO_4$): an irritant; moderate health rating
- **Sodium hydroxide** (NaOH): toxic, corrosive and an irritant; severe health rating
- **Water** [distilled]

Protection
Wear goggles, gloves and protective clothing when handling NaOH.

Other tests to consider or confirm results

- Test for organic sulfur using calcium oxalate and pyrolysis (p. 148).
- Test for sulfur using lead acetate paper and pyrolysis (p. 146).
- Test for protein (nitrogen) using calcium oxide and pyrolysis (p. 142).

Reference
Browning, B. L. 1969. *Analysis of Paper.* New York: Marcel Decker.

Reagent preparation

- **2% aqueous copper(II) sulfate ($CuSO_4$) solution:** add 0.5g $CuSO_4$ to 25mL distilled water.
- **5% 1.2M aqueous sodium hydroxide (NaOH) solution (w/v):** add 1.25g NaOH pellets to 25mL distilled water. ALWAYS ADD CAUSTICS TO WATER!

Method of sampling

An inconspicuous area of the paper should be tested or a few paper fibers may be adequate. For other materials a small sample must be removed for testing.

Procedure

For paper
1. Place a drop of the copper(II) sulfate solution on the paper.
2. Wait a few minutes.
3. Remove excess copper(II) sulfate solution with blotting paper.
4. Place a drop of the 5% NaOH solution on the same spot on the paper.
5. Look for color formation.

For other proteinaceous materials
1. Place a small piece of sample or a few grains of ground sample on a spot-test plate.
2. Add a drop of copper(II) sulfate solution to the sample.
3. Wait a few minutes. The sample should absorb some of the solution and turn slightly bluish.
4. Remove the excess copper(II) sulfate solution by wicking it onto a piece of blotting paper.
5. Add a drop of the NaOH solution to the sample.
6. Look for color formation.

Observations and interpretation

The formation of a purple color indicates the presence of protein, the test is not sensitive to trace quantities. Sometimes the reaction takes up to an hour to show a positive. Ignore any blue color as it comes from the reaction of sodium hydroxide with copper(II) sulfate. If the sample is small, e.g. a protein binder, it may be necessary to observe the specimen under magnification. If the sample has a cutinaceous layer such as quills or certain hairs, the purple will only show up along the cut edges. It may be necessary to soak organic materials such as horn, tortoiseshell, sinew, etc. for an hour and then absorb excess solution with blotting paper before the sodium hydroxide is added.

Storage and reagent shelf life

Copper(II) sulfate is stable when stored in closed containers in ambient conditions. Sodium hydroxide is stable when stored in sealed containers in ambient conditions, but is very hygroscopic. It will absorb carbon dioxide from the air and become less basic if the container is not sealed tightly. The reagent solutions are stable when stored in sealed containers in ambient conditions.

Test for sulfur using lead acetate paper and pyrolysis

Purpose
To determine if an object contains sulfur, which is an indication that it is made of rubber or ebonite. The test may also indicate the presence of sulfur in protein-based materials such as hair, leather and hide glues.

Principle
Pyrolysis of the sample causes the formation of sulfides which react with lead acetate to form brown-black lead sulfide.

$$\text{sulfur containing compound} \xrightarrow{\text{heat}} \underset{\text{organic thiol or sulfide}}{\text{RSH or RSR}}$$

$$(R = \text{carbon containing chain})$$

$$\text{RSH} + \text{Pb}(CH_3COO)_2 \rightarrow \text{PbS}\ (s)\ \text{or}\ \text{Pb}(RS)_2\ (s)$$

$$\text{(brown-black)}$$

Equipment
- Pasteur pipettes
- laboratory wrapping film
- pH indicator paper
- source of flame (alcohol lamp, Bunsen burner)

Reagents and safety
- **3% hydrogen peroxide (H_2O_2)**: an irritant; slight health rating
- **Lead acetate test paper ($Pb(C_2H_3O_2)_2$)** [Macherey-Nagel]: highly toxic, carcinogen and a mutagen; severe health rating
- **Water** [distilled]

Protection
Wear gloves when handling the papers. Open flames in a conservation lab are always a potential hazard. Extreme caution should be used when using any type of open flame. There should be no solvent containers in the vicinity or any other materials that can catch fire. Wear eye protection when performing this test. The fumes from the burning plastic should not be inhaled – perform test in well-ventilated area or fume hood if possible.

Other tests to consider or confirm results
- Test for organic sulfur using calcium oxalate and pyrolysis (p. 148).
- Tests for proteins (pp. 144, 148).

Reference
Rémillard, France. 1995. *La Preservation des Objects de Plastique et de Caoutchouc. Seminaire De Formation*. Quebec: Centre de Conservation du Quebec.

Reagent preparation
None.

Method of sampling
A small sample must be removed from the object.

Procedure

1. Seal the tip of a Pasteur pipette with a flame source.
2. Introduce the sample into the pipette and push it down into the tapered area.
3. Cut a strip of lead acetate paper, about 2cm in length.
4. Fold the strip along its long axis and quickly dip one end into distilled water.
5. Place the lead acetate paper (wet end towards sample) just inside the open end of the pipette.
6. Seal the opening of the pipette with laboratory sealing film (do not allow it to touch the test paper).
7. Gently heat the sample in the flame until fumes evolve.
8. Allow enough time for the fumes to reach the test paper.
9. Observe any color change in the paper.

Observations and interpretation
The formation of a brown color on the lead acetate test paper confirms the presence of sulfur. Additional confirmation may be obtained by treating the paper with hydrogen peroxide solution. This removes the color by converting lead sulfide to lead sulfate, which is white. The addition of a pH paper strip in the reaction chamber (pipette) allows one to determine the pH of the fumes. If the test is negative for sulfur but the fumes are highly acidic, the material may contain poly(vinyl chloride), which can be confirmed with the Beilstein test.

Storage and reagent shelf life
The test papers are stable if stored in a cool dry place. Hydrogen peroxide is stable in ambient conditions, avoid light and protect from freezing.

Test for organic sulfur using calcium oxalate and pyrolysis

Purpose
To determine the presence of organically bound sulfur in fibers such as wool or silk.

Principle
Sulfur compounds in the fibers are decomposed to H_2S during pyrolysis in the presence of calcium oxalate. The gas released reacts with lead acetate paper forming lead sulfide that turns the paper a darker color.

$$\text{sulfur-containing compound} + Ca^{2+} \left[O-\underset{\underset{calcium\ oxalate}{}}{\overset{\overset{O}{\|}}{C}} - \overset{\overset{O}{\|}}{C} - O \right]^{2-} \longrightarrow H_2S\ (g) + CaCO_3\ (s) + CO_2\ (g)$$

$$H_2S\ (g) + Pb^{2+}\ (aq) \rightarrow PbS\ (s) + 2H^+\ (aq)$$
$$\text{(brown)}$$

Equipment

- test tube
- test-tube holder
- microspatula
- scissors
- source of flame (Bunsen, or alcohol)
- laboratory wrapping film
- Pasteur pipette

Reagents and safety

- **Calcium oxalate** (CaC_2O_4): an irritant; slight health rating
- **Lead acetate paper** ($Pb(C_2H_3O_2)_2$): highly toxic, carcinogen and a mutagen; severe health rating

Protection
Wear gloves when handling the lead acetate papers. Open flames in a conservation lab are always a potential hazard. Extreme caution should be used when using any type of open flame. There should be no solvent containers in the vicinity or any other materials that can catch fire. Wear eye protection when performing this test. The fumes from the decomposing fibres should not be inhaled – perform the test in a well-ventilated area or fume hood if possible.

Other tests to consider or confirm tesults

- Test for sulfur using lead acetate paper and pyrolysis (p. 146).
- Test for protein using copper(II) sulfate (Biuret test) (p. 144).
- Test for protein (nitrogen) using calcium oxide and pyrolysis (p. 142).

Reference
Feigl, Fritz, and Vinzenz Anger. 1972. *Spot Tests in Inorganic Analysis*, 6th English edition. New York: Elsevier.

Reagent preparation
None.

Method of sampling
A small sample must be removed.

Procedure
1. A small sample is inserted into a test tube. (This test can also be performed by the Rémillard pipette technique.)
2. A microspatula tip full of calcium oxalate is added.
3. A small piece of the lead acetate paper is folded in half and inserted into the top of the test tube.
4. The test tube is held in the test-tube holder and heated in a flame until the sample burns.

Observations and interpretation
If organic sulfur is present in the sample, the paper will turn brown. The more sulfur that is present, the darker the paper will turn. Most fibers that contain wool or silk will cause the paper to darken to resemble a coffee-like stain. During testing, a yarn sample made from 5% wool showed distinct darkening of the test paper. The test should also be performed on a sample of known sulfur content, such as a piece of wool, so that the color change can be compared with the unknown.

Storage and reagent shelf life
The lead acetate papers are stable in ambient conditions. Calcium oxalate is stable when stored in a tightly closed container in ambient conditions.

Test for phenols in vegetable-tanned leather using lead acetate

Purpose
To determine the presence of vegetable tannins in leather.

Principle
Because of the presence of phenolic hydroxyl groups in the tannins, salt complexes are formed with many metals. When exposed to lead acetate, an extracted solution of tannin salts will form a precipitate.

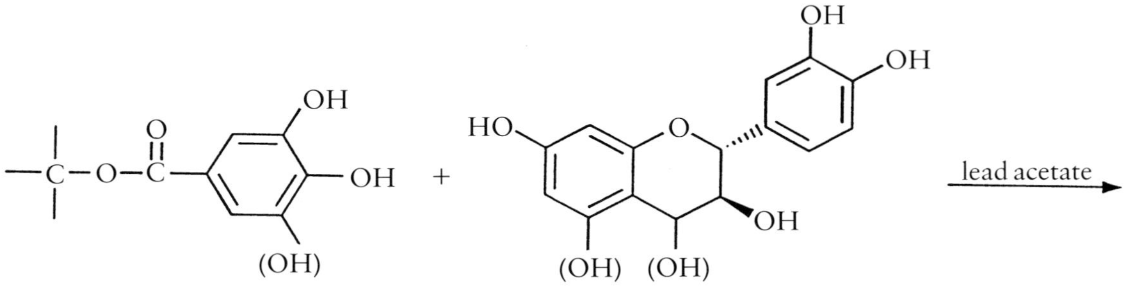

polyphenolic ester
(hydrolizable tannins)

flavonoid compounds
(condensed tannins)

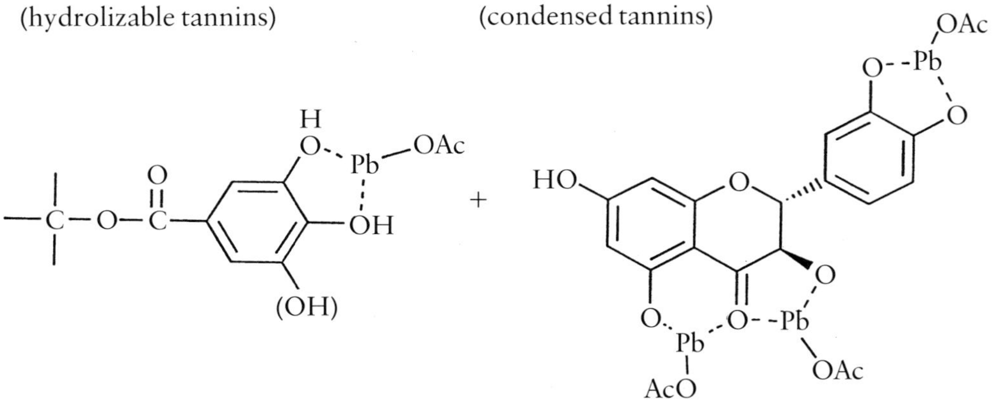

three possible complexation sites
(only one is necessary to impart color)

Equipment

- graduated pipette
- funnel
- test tubes with stoppers
- filter paper
- scalpel
- balance (to weigh 0.1g)

Reagent and safety

- **Acetone:** flammable, irritant; slight health rating
- **Lead acetate** $Pb(CH_3COO)_2$: toxic, carcinogen and a mutagen; severe health rating
- **Water** [distilled]

Protection
Wear goggles, gloves, and protective clothing when handling lead acetate.

Other tests to consider or confirm results

- Test to determine vegetable-tanned leather using iron(III) sulfate (p. 152).

Reference
Reed, R. 1972. *Ancient Skins, Parchments and Leathers*. London: Seminar Press.

Reagent preparation

- **50% aqueous acetone solution:** add 1mL acetone to 1 mL distilled water.
- **2% aqueous lead acetate Pb(CH$_3$COO)$_2$ solution:** add 0.5g Pb(CH$_3$COO)$_2$ to 25mL distilled water.

Method of sampling

Remove a small sample for testing with a scalpel; a 0.01–0.02g sample should be sufficient.

Procedure

1. Place sample in a test-tube with 2mL 50% aqueous acetone solution for four hours.
2. Filter off the extract using a funnel and filter paper into a test-tube.
3. Measure 1mL of the extract solution using a graduated pipette and pour into a clean, stoppered test tube.
4. Measure 1mL of the 2% lead acetate reagent and pour it into the test tube with the 1mL of extract.
5. Observe for precipitate formation.
6. Allow the solution to stand for a few minutes until the precipitate settles out.

Observations and interpretation

A dense light brown precipitate that decolorizes the liquid indicates the presence of vegetable tannins.

Test to determine vegetable-tanned leather using iron(III) sulfate

Purpose
To determine the presence of vegetable tannins in leather.

Principle
Iron(III) (ferric) ions react with the phenolic compounds present in vegetable-tanned leathers, giving a dark blue or green coloration.

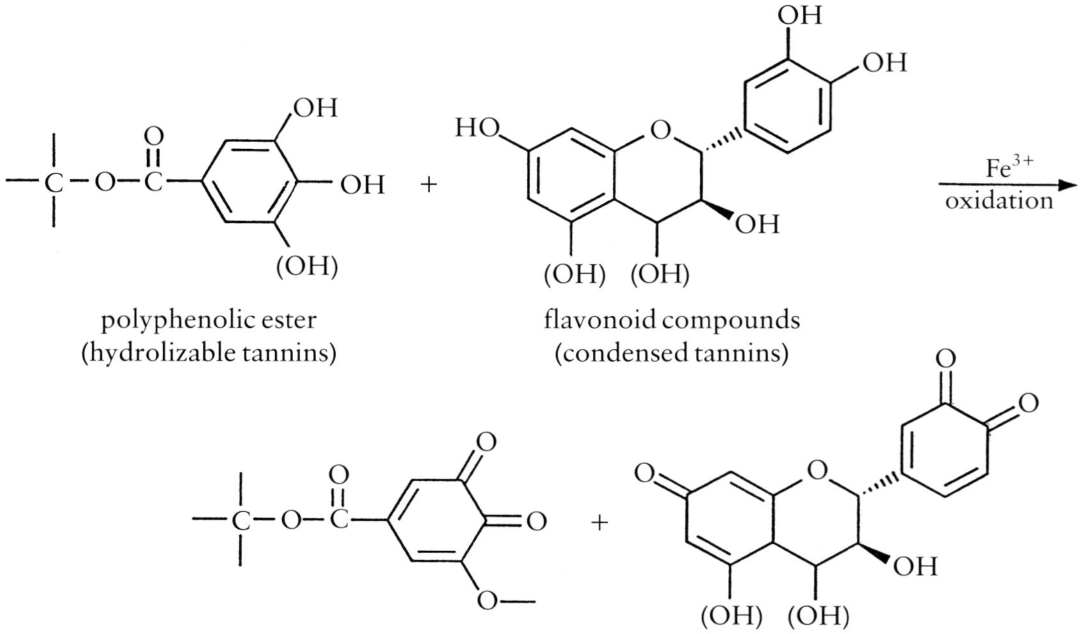

polyphenolic ester
(hydrolizable tannins)

flavonoid compounds
(condensed tannins)

$\xrightarrow{Fe^{3+} \text{ oxidation}}$

hydroxyquinone and ortho-quinone derivatives
(green-blue)

Equipment

- Pasteur pipettes
- glass slides
- glass slide covers
- reagent container
- tweezers
- balance (to weigh 0.1g)
- magnification (optional)

Reagents and safety

- **Iron(III) sulfate** (ferric sulfate, $Fe_2(SO_4)_3$): an irritant; slight health rating
- **Water** (distilled)

Other tests to consider or confirm results

- Test for phenols in vegetable-tanned leather using lead acetate (p. 150).

Reference
Florian, Mary-Lou. 1984. Analytical Methods for Protein Identification and Characterization; Simple Analytical Methods Using Case Studies to Illustrate Techniques, Section 5. In *Protein Chemistry for Conservators*, C. L. Rose and D. W. Von Endt, editors, 89–96. Washington, DC: American Institute for Conservation.

Reagent preparation

- **2% iron(III) sulfate solution**: add 0.5 g $Fe_2(SO_4)_3$ to 25mL distilled water. *Note:* iron(III) sulfate can be difficult to dissolve, so agitation or the use of a magnetic stirrer will be helpful. If allowed to stand after thorough stirring, it will fully dissolve into a transparent yellow solution.

Method of sampling
Remove two small samples of leather (a few fibers each).

Procedure

1. Place two samples on a single glass slide and cover each with separate cover slips.
2. Using a Pasteur pipette, place a drop of distilled water at the edge of the cover slip of one sample (a control). Dampening the control fiber may afford a more accurate comparison with the reagent-wetted sample.
3. Using a Pasteur pipette place a drop of the iron(III) sulfate solution at the edge of the cover slip of the test sample.
4. Observe the color change in the treated sample by comparing it to the dampened control sample adjacent to it.

Observations and interpretation
A dark blue or green color indicates the presence of phenols, hence the presence of vegetable tannins. Observation with low power magnification may be helpful. Some authors have reported that a blue-gray coloration indicates the presence of hydrolyzed tannins (polysaccharides) and a green-gray coloration indicates the presence of condensed tannins (lignin-like polycyclic hydrocarbons), but these subtle color variations can be very difficult to distinguish.

Storage and reagent shelf life
The iron(III) sulfate solution is stable when stored in sealed dark containers in ambient conditions; protect from light.

Test for indigo using sodium hydrogen sulfite

Purpose
To determine the presence of indigo in blue textiles, paints, or pigments.

Principle
The dye in the material being tested is reduced to the leuco form of the dye and then extracted from its aqueous solution with ethyl acetate; the extract (if it is indigo) colors the ethyl acetate a bright blue.

indigo + NaHSO$_3$ (aq) $\xrightarrow{\text{base}}$ leuco indigo, or indigo white (blue color in ethyl acetate on exposure to air)

Equipment

- stoppered micro-test tubes 10mL (larger ones will work)
- hot plate
- beaker
- balance (to weigh 1.g)
- droppers
- reagent container

Reagents and safety

- **Sodium hydroxide** (NaOH): toxic, corrosive and irritant; severe health rating
- **Sodium hydrogen sulfite** (sodium bisulfite) (NaHSO$_3$): irritant, and flammable; moderate health rating
- **Ethyl acetate** (CH$_3$COOCH$_2$CH$_3$): irritant and flammable; moderate health rating
- **Water** [distilled]

Protection
Wear goggles, gloves and protective clothing when handling the reducing solution.

Other tests to consider or confirm results
None.

Reference
Hofenk-de Graaff, Judith H. 1974. A Simple Method for the Identification of Indigo. *Studies in Conservation* **19**(1): 54–5.

Reagent preparation

- **Reducing solution**: dissolve 1.25g sodium hydroxide (NaOH) and 1.25g sodium hydrogen sulfite ($NaHSO_3$) in 25mL distilled water. Keep tightly stoppered.

Method of sampling

The dye must be in solution for the test to succeed. If the sample is a textile, it may be necessary to heat a small sample together with sodium hydrogen sulfite solution in a test tube in a water bath to extract some of the dye.

Procedure

1. Place a small sample of the dye solution in a test tube.
2. Add three to five drops of the reducing solution.
3. Stopper the test tube. The reducing agent smells so it should be kept covered at all times.
4. Wait until the dye turns yellow (i.e. is reduced by the reducing solution).
5. Add three to five drops of ethyl acetate.
6. Stopper the test tube and shake. The solution will separate into two distinct layers.
7. Hold the test tube against a white background and look for a color change.

Observations and interpretation

If indigo is present, the upper layer will turn blue on oxidation. The test is sensitive to 5mg of indigo but for tiny samples the change to the blue color may be difficult to see. No other natural blue or violet dyes give the same reaction, although some modern synthetics do.

Storage and reagent shelf life

Ethyl acetate is stable when stored in a sealed container in ambient conditions. The reducing solution oxidizes readily and should be prepared each time.

Test for lignin using phloroglucinol

Purpose
To determine the presence of lignin in a material such as paper, fiber or wood. This test may be useful in distinguishing wood from leaf elements in basketry and degraded wood from bone in an archaeological context.

Principle
A solution of phloroglucinol (1,3,5-benzenetriol) stains groundwood and other lignin-containing fibers in proportion to the amount present. Lignin is present in most woods, woody plant material and in paper made from wood fibers.

lignin fragment + phloroglucinol $\xrightarrow{acid}$ (red-violet)

Equipment
- droppers
- acid-resistant reagent container
- balance (to weigh 0.01g)
- graduated cylinder
- magnification (optional)

Reagents and safety
- **Phloroglucinol** ($C_6H_3(OH)_3 \cdot 2H_2O$): irritant; slight health rating
- **Hydrochloric acid** (HCl): toxic and corrosive; severe health rating
- **Methanol** (methyl alcohol): toxic, flammable and irritant; severe health rating
- **Water** [distilled]

Protection
Wear goggles, gloves and protective clothing when handling HCl and methanol.

Other tests to consider or confirm results
- Test for cellulose using aniline acetate and pyrolysis (p. 160).
- Test for cellulose and its derivatives using 1-naphthol (Molisch test) (p. 162).

Reference
Browning, B. L. 1969. *Analysis of Paper*. New York: Marcel Dekker.

Reagent preparation

- **Phloroglucinol solution (approx. 5%, w/v):** dissolve 0.25g phloroglucinol in a mixture of 15mL methanol, 15mL distilled water, followed by 15mL concentrated hydrochloric acid (HCl).

Method of sampling

The test is carried out directly on a small piece of paper, on individual fibers or on a few particles of the material.

Procedure

1. Add one drop of phloroglucinol solution to the sample.
2. Wait one minute.

Observations and interpretation

If lignin is present in large amounts a strong magenta (beet red) or violet color develops. If small amounts of lignin are present a weak stain is formed and examination under low power magnification is recommended.

Storage and reagent shelf life

The phloroglucinol solution turns yellow when exposed to light and when stored for long periods. Store in a dark bottle or cover with aluminum foil to protect from light.

Test for rosin using sulfuric acid (Raspail test)

Purpose
To determine the presence of rosin in objects such as accretions on baskets or amber (fossilized resin). This test has been suggested in paper conservation to determine if paper has been sized with rosin.

Principle
In the presence of sugar (sucrose), rosin becomes red when exposed to concentrated sulfuric acid (H_2SO_4).

$$\text{rosin} + \text{sugar} \xrightarrow{\text{acid}} \text{red color}$$

Equipment

- spot-test plate
- droppers
- frosted microscope slide
- blotting paper
- acid-resistant reagent containers

Reagents and safety

- **Sugar** (table sugar will do)
- **Sulfuric acid** (H_2SO_4): toxic, oxidizing and corrosive; severe health rating
- **Water** [distilled]

Protection
Wear gloves, goggles and protective clothing while handling H_2SO_4.

Other tests to consider or confirm results
None.

Reference
Browning, B. L. 1969. *Analysis of Paper.* New York: Marcel Dekker.

Reagent preparation
- **Saturated sugar solution:** add sufficient sugar (approximately 35–40g) to 25mL of distilled water so that there are some crystals that will not dissolve in the flask. It may be necessary to leave the solution overnight to make sure that it is saturated.

Method of sampling
A small, thin sample must be removed for testing.

Procedure
For solid samples
1. Scratch a small amount of sample onto a frosted microscope slide or place the sample on the spot-test plate.
2. Apply a drop of the sugar solution to the sample.
3. After about 10 seconds remove the excess sugar solution with a blotter or alternatively move the sample to another spot-test well.
   Note: Test results are improved if all excess sugar is removed.
4. Apply one small drop of concentrated sulfuric acid (H_2SO_4) to the sample.
5. Wait one minute for the color to develop.

For paper
1. Place a drop of the sugar solution directly onto the paper sample.
2. Wait for about 10 seconds and remove any excess sugar solution with a blotter.
3. Apply one small drop of concentrated sulfuric acid (H_2SO_4) to the test spot.
4. Wait one minute for the color to develop.

Observations and interpretation
The development of a raspberry red color after a minute indicates the presence of rosin. The reaction is easier to see with tiny sample amounts. If there is much fresh rosin in the material the color will start to develop immediately. The color will start to bleed into the surrounding drop of sulfuric acid. In materials where the rosin content is low or in amber (fossilized resin), the raspberry red may not develop until after 10–30 minutes have passed. It is useful to run a blank control by placing a drop of sulfuric acid on a sample of the material without first putting it in the sugar solution. If it is not rosin, the material will appear the same in sulfuric acid with or without the addition of the sugar solution. Many materials turn brown or red when in contact with sulfuric acid. Also, if there is too much sugar solution, a brown color may develop due to the reaction of the sugar with the sulfuric acid. The true positive color for rosin is a raspberry red.

Storage and reagent shelf life
Sulfuric acid is stable when stored in a sealed, acid-resistant container in ambient conditions. The sugar solution is stable in ambient conditions.

Test for cellulose using aniline acetate and pyrolysis

Purpose
To determine the presence of cellulose in an object such as paper, plant material, adhesive or plastic (cellulose nitrate, cellulose acetate).

Principle
Pyrolysis in the presence of ortho-phosphoric acid (H_3PO_4) causes the formation of gases that react with aniline acetate to give a pink or yellow (cellulose nitrate only) spot.

$$\text{cellulose} \xrightarrow{acid} \text{simple sugars} \xrightarrow{acid} \text{furfural fragment}$$

$$\text{furfural} + 2\,\text{aniline acetate} \xrightarrow{heat} \text{(pink product)}$$

Equipment

- source of flame (Bunsen or alcohol burner)
- test tube
- filter paper
- graduated cylinder
- acid-resistant reagent container
- dropper

Reagents and safety

- **Phosphoric acid** (H_3PO_4): corrosive and irritant; moderate health rating
- **Aniline** ($C_6H_5NH_2$): toxic, a carcinogen, flammable and irritant; severe health rating
- **Glacial acetic acid** (CH_3COOH): toxic, corrosive and flammable; moderate health rating
- **Water** [distilled]

Protection
Wear goggles, gloves and protective clothing when handling aniline and acids. Avoid contact or breathing fumes. Open flames in a conservation lab are always a potential hazard. Extreme caution should be used when using any type of open flame. There should be no solvent containers in the vicinity or any other materials that can catch fire. Wear eye protection when performing this test. The fumes from the decomposing sample should not be inhaled. Perform test in well-ventilated area or fume hood if possible. Also note that the sample is likely to melt on to the bottom of the test tube which renders it useless for further work.

Other tests to consider or confirm results
- Test for cellulose and its derivatives using 1-naphthol (Molisch test) (p. 162).
- Test for nitrate (cellulose nitrate) using diphenylamine (p. 164).

Reference
Browning, B. L. 1969. *Analysis of Paper*. New York: Marcel Dekker.

Reagent preparation

- **Aniline acetate solution**: prepare a 1:1 (8.5M) solution of glacial acetic acid and distilled water. Add this solution to the aniline until cloudiness disappears (for example, approximately 10mL of the dilute acetic acid is necessary with 10mL aniline).

Method of sampling
A small sample must be removed for pyrolysis.

Procedure

1. Place a small sample of the material in a test tube.
2. Add one drop concentrated phosphoric acid.
3. Cover the test tube with a piece of filter paper.
4. Place a drop of aniline acetate solution on the filter paper.
5. Heat the test tube with the flame until the sample is carbonized.

Observations and interpretation
If the material is cellulose or a nearest derivative, a pink spot will appear on the filter paper. Cellulose nitrate forms a yellow spot. *Caution:* Cellulose nitrate can spontaneously ignite when exposed to heat.

Storage and reagent shelf life
Aniline is stable in ambient conditions. The aniline acetate solution should be stable for several weeks if stored in a sealed container in ambient conditions. Glacial acetic acid is stable when stored in a sealed, acid-resistant container in ambient conditions. Phosphoric acid is stable when stored in a sealed, acid-resistant container in ambient conditions.

Test for cellulose and its derivatives using 1-naphthol (Molisch test)

Purpose
To identify cellulose esters such as cellulose acetate or cellulose nitrate, cellulose ethers such as methyl cellulose or ethyl cellulose, and other regenerated cellulosics such as viscose (rayon). This test will also identify cellulose in objects such as paper, wood, cotton and linen.

Principle
Cellulose is a polysaccharide, and when it and its derivatives are treated with concentrated acid, they hydrolyze into glucose, which forms a red or purple-colored complex in the presence of 1-naphthol.

hydrolyzed cellulose and derivatives (cellulose acetate and cellulose nitrate) + 1-naphthol $\xrightarrow{acid}$ red complex (cellulose acetate) + yellow complex (cellulose nitrate)

Equipment
- balance (to weigh 1g)
- low magnification (×10) (optional)
- droppers
- microscope slide cover slips
- microscope slides with frosted ends
- container

Reagents and safety
- **Ethanol** [ethyl alcohol]: flammable and an irritant; slight health rating
- **1-naphthol** [also known as α-naphthol]: irritant, toxic, and oxidizing; severe health rating
- **Sulfuric acid** (H_2SO_4): toxic, oxidizing, and corrosive; severe health rating

Protection
Wear goggles, gloves, and protective clothing when handling H_2SO_4.

Other tests to consider or confirm results
- Test for cellulose using aniline acetate and pyrolysis (p. 160).
- Test for nitrate (cellulose nitrate) using diphenylamine (if testing for cellulose nitrate) (p. 164).
- Test for polyester groups using hydroxylamine hydrochloride (if testing for cellulose esters) (p. 168).

Reference
Rémillard, France. 1995. *La Preservation des Objects de Plastique et de Caoutchouc. Seminaire de Formation.* Quebec: Centre de Conservation du Quebec.

Reagent preparation

- **2% (w/v) 1-naphthol solution**: add 0.5g of 1-naphthol to 25mL of ethanol.

Method of sampling

A sample is removed by rubbing the object against the frosted end of a microscope slide, or a small sample from the object may be removed. The sample must be sufficiently small to allow a microscope cover slip to lie flat on top of it.

Procedure

1. Place or deposit a small amount of sample on the microscope slide and cover with cover slip.
2. Add several drops of the 2% 1-naphthol solution to wick into the sample.
3. Allow the ethanol to evaporate.
4. Place one to two drops of concentrated sulfuric acid next to the sample and allow the drops to wick towards the sample and flow past it.
5. Observe the sample for a color change over a 5–15 minute interval; magnification and a white background may help in this observation.

Observations and interpretation

As cellulose (or its derivatives) are hydrolyzed (broken down) into glucose by the acid, the sample will often progress through a series of colors (from yellow to brown to red and finally to purple), but in a positive reaction a red or deep purple color appears after 10–15 minutes. The cover slip helps in the observation as it keeps the colored complex in a thin, relatively transparent layer. A green color indicates the presence of cellulose nitrate. The 1-naphtol reagent dries forming small crystals that will turn yellow when the acid is applied. Thus, a yellow color that does not change after 15 minutes should be considered a negative result. Lignin and non-cellulosic carbohydrates may cause the sample to turn brown or black. Very decayed cellulose nitrate gives a negative reaction.

Storage and reagent shelf life

The 1-naphthol solution may be kept in a sealed container for about a month in ambient conditions. Sulfuric acid (H_2SO_4) is stable when stored in a sealed acid-resistant container in ambient conditions.

Test for nitrate (cellulose nitrate) using diphenylamine

Purpose
To determine the presence of nitrates such as cellulose nitrate plastic, film or adhesive. The test also works with nitrate salts (sodium nitrate, silver nitrate etc.) but will not work in the presence of reduced nitrogen such as is found in proteins or ammonia-containing compounds.

Principle
Sulfuric acid reacts with cellulose nitrate to form nitronium ions (NO_2^+) which then react with diphenylamine (DPA) forming diphenylbenzidine violet.

$$\text{cellulose nitrate} + H_2SO_4 \rightarrow NO_2^+$$
nitronium ion
(oxidizing agent)

$$2 \text{ diphenylamine (DPA)} \xrightarrow{2NO_2^+,\ -2H^+,\ 2NO_2(g)} \text{diphenylbenzidine – DPB}$$

$$\text{DPB} \xrightarrow{2NO_2^+,\ -2H^+,\ 2NO_2(g)} \text{diphenylbenzidine violet – DPBV (dark blue)}$$

Equipment
- spot-test plate
- droppers
- balance (to weigh 0.1g)
- acid-resistant reagent containers
- graduated cylinders
- microspatula
- scalpel
- glass stirring rods
- microscope slides with frosted ends

Reagents and safety
- **Diphenylamine** ((C_6H_5)$_2$NH): irritant; slight health rating
- **Sulfuric acid** (H_2SO_4) toxic, oxidizing and corrosive; severe health rating
- **Water** [distilled]

Protection
Wear goggles, rubber gloves and protective clothing when first making up the solution.

Other tests to consider or confirm results
- Test for cellulose using aniline acetate and pyrolysis (p. 160).
- Test for nitrate using spot-test papers (p. 112).
- Test for nitrate using iron(II) sulfate (p. 114).

Reference
Williams, R. Scott, 1989. *The Diphenylamine Spot Test for Cellulose Nitrate in Museum Objects. CCI Notes* No. 17/2. Ottawa: Canadian Conservation Institute.

Reagent preparation

- **Diphenylamine solution**: slowly add 0.9mL concentrated sulfuric acid (H_2SO_4) to 1mL distilled water with constant stirring (heat is generated); add 0.05g of diphenylamine (($C_6H_5)_2NH$) in small increments. Store in an acid-resistant storage bottle. ALWAYS ADD ACID TO WATER!

Alternative preparations
- Add 20mg diphenylamine to 1mL of concentrated sulfuric acid (H_2SO_4) (Koob 1982).
- As a last resort (in the event that a sensitive balance is not available that can measure milligrams) fill the tip of a no. 15 scalpel blade with diphenylamine (($C_6H_5)_2NH$) crystals (making sure the blade is covered up to the point where the cutting edge stops). Add 20 drops of concentrated sulfuric acid (H_2SO_4).

Method of sampling

A small chip or scraping is removed from an inconspicuous place of an object and subjected to the test or it may be rubbed onto the frosted end of a microscope slide; if a low power microscope is utilized the sample used may be very small.

Procedure

1. Scratch a small amount of sample onto a frosted microscope slide or place a small piece of the sample on the spot-test plate.
2. Using a dropper, place one drop of diphenylamine/sulfuric acid solution on the sample.
3. Observe for color change.

Observations and interpretation

If cellulose nitrate is present, a dark blue color will develop and spread out in the reagent drop; the development of no color or yellow, green, orange or brown colors are considered negative results. Sometimes the color at first appears black and then gradually becomes dark blue as the color is diluted in the drop of acid. With very old cellulose nitrate (such as nitrate-based photographic negatives) the color may take as long as 30 seconds to appear. It is useful to run a control simultaneously using a known sample of cellulose nitrate (old adhesive or plastic). This will provide a good color indicator for comparison.

Storage and reagent shelf life

Sulfuric acid (H_2SO_4) is stable when stored in ambient conditions in a sealed acid-resistant container away from heat and sunlight. Diphenylamine is stable if protected from light in ambient conditions but it will discolor on exposure to light. If stored in a proper container the reagent is stable for more than a year, even if it becomes slightly dark in color.

Test for poly(vinyl alcohol) using iodine/potassium iodide

Purpose
To determine the presence of poly(vinyl alcohol) [PVA also referred to as PVOH], or one of it derivatives such as poly(vinyl acetate) [PVAC or PVAc], or a poly(vinyl ketal) [Alvar and Butvar].

Principle
The test solution reacts with PVOH or one of its derivatives to form color.

$$I_2\ (s) + I^-\ (aq) \rightarrow I_3^-\ (aq)$$

triiodide ion

$$I_3^-\ (aq) + \text{poly(vinyl alcohol)} \rightarrow \text{colored complex}$$

(or derivative)

Equipment

- droppers
- balance (to weigh 0.1 g)
- graduated cylinder
- filter paper
- microspatula
- acid-resistant reagent containers
- spot-test plate
- magnification (optional)
- beaker

Reagents and safety

- **Acetone**: flammable and irritant; slight health rating
- **Glacial acetic acid**: corrosive, toxic and flammable; moderate health rating
- **Glycerol** (glycerine): irritant; slight health rating
- **Iodine** (I_2): toxic and corrosive; moderate health rating
- **Potassium iodide** (KI): irritant; moderate health rating
- **Water** [distilled]

Protection
Wear goggles, gloves and protective clothing when handling acids.

Other tests to consider or confirm results
None.

Reference
Howie, Francis M. P. 1984. 'Materials used for conserving fossil specimens since 1930: a review' in *Adhesives and Consolidants*, N.S. Brommelle, Elizabeth M. Pye, Perry Smith, Garry Thompson, editors. 92–7. London: International Institute for Conservation, Paris Congress, preprint.

Reagent preparation

- **Solution A**: dissolve 0.5g potassium iodide (KI) in 20mL distilled water, add 0.45g iodine (I_2) and after both are dissolved add 1g glycerol. This solution should appear dark brown.
- **Solution B**: add equal parts of glacial acetic acid to distilled water. ALWAYS ADD ACID TO WATER!

Method of sampling
A small sample must be removed for testing.

Procedure

1. Place a small sample on a spot-test plate and add two drops of acetone and allow the solvent to evaporate.
2. Immediately prior to use, add seven parts of solution A to 10 parts solution B in a beaker.
3. Place a drop of the mixed solution on the sample.
4. Wait at least 10 minutes (or longer) and observe the edges of the sample.

Observations and interpretation
If the sample is a PVOH-derivative it will form color, which may be green, purple, blue, red, yellow, or in-between. Acrylics and polyesters do not form color but may absorb the yellow of the iodine solution and appear dark yellow or brown (epoxies may appear yellow with a red edge); shellac will turn pale yellow, but its presence can be confirmed independently.

Storage and reagent shelf life
Acetone is stable in ambient conditions. Glacial acetic acid is stable in ambient conditions, but heat and sunlight can contribute to instability, it should be stored above 17°C or 63°F. Glycerol is stable in ambient conditions. Iodine is stable in ambient conditions, but should be stored in a dark container. Potassium iodide is stable in ambient conditions, but with long exposure to air it becomes yellow, due the release of iodine. Solutions A and B are stable if stored separately in sealed containers at room temperature.

Test for polyester groups using hydroxylamine hydrochloride

Purpose
To determine the presence of esters, such as polyester in plastics, binders, resin, cloth, thread, and film.

Principle
The reaction of esters with hydroxylamine hydrochloride ($NH_2OH \cdot HCl$) forms a hydroxamic acid. This hydroxamic acid reacts with hydrochloric acid and ferric chloride to form a purplish complex.

$$\left[-O-(CH_2)_n-O-\overset{O}{\underset{\|}{C}}-(CH_2)_n-\overset{O}{\underset{\|}{C}}- \right] + NH_2OH \cdot HCl \xrightarrow{\text{base}} -\overset{O}{\underset{\|}{C}}-\overset{H}{\underset{}{N}}-OH$$

a polyester fragment   a hydroxamic acid

$$\text{hydroaxamic acids} + Fe^{3+}\,(aq) \rightarrow \text{violet complex}$$

Equipment

- test tube
- droppers
- hot plate
- beaker
- microspatula
- acid-resistant reagent containers
- balance (to weigh 0.1g)
- graduated cylinder
- thermometer

Reagents and safety

- **Hydrochloric acid** (HCl): toxic and corrosive; severe health rating
- **Hydroxylamine hydrochloride** ($NH_2OH \cdot HCl$) [oxammonium hydrochloride]: toxic, corrosive and a possible mutagen; severe health rating
- **Iron(III) chloride** ($FeCl_3$) [ferric chloride]: corrosive; moderate health rating
- **Methanol** (methyl alcohol): toxic, flammable and irritant; severe health rating
- **Potassium hydroxide** (KOH): toxic and corrosive; severe health rating
- **Water** [distilled]

Protection
Wear gloves, goggles and protective clothing when handling these reagents.

Other tests to consider or confirm results
None.

Reference
Rémillard, France. 1995. *La Preservation des Objects de Plastique et de Caoutchouc. Seminaire de Formation.* Quebec: Centre de Conservation du Quebec.

Reagent preparation

- **11% w/w methanolic potassium hydroxide (KOH) solution** (also known as methanolic KOH or alcoholic caustic potash): dissolve 5.5g KOH in 44.5g methanol.
- **10% w/w methanolic hydroxylamine hydrochloride (HH) solution**: dissolve 5g hydroxylamine hydrochloride ($NH_2OH \cdot HCl$) in 45g methanol.
- **Saturated iron(III) chloride ($FeCl_3$) solution in water**: add sufficient $FeCl_3$ to distilled water so that some crystals remain undissolved. $FeCl_3$ is very soluble in water but it takes some time to dissolve (for example, adding approximately 1.0g $FeCl_3$ to 1mL water is adequate).
- **4M hydrochloric acid (HCl) solution (1:2)**: add one part conc. HCl to two parts distilled water. The concentration of the acid is important in this test, if a more concentrated solution of hydrochloric acid is used, the reaction will appear dark brown. **ALWAYS ADD ACID TO WATER!**

Method of sampling

A small sample of the material must be removed for testing. For threads, the sample should be at least 3cm long.

Procedure

1. Place a small sample in a test tube.
2. Add ten drops of the KOH solution.
3. Add six drops of the HH solution. The solution may turn white when the KOH and HH solutions are mixed. This does not affect the test.
4. Heat test tube in the water bath (temperature 50°C±2°C (120°F) for two minutes. Keeping the temperature within this range may be important for the success of this test so that the methanol does not boil off.
5. Remove test tube from bath.
6. Add ten drops of the HCl solution.
7. Add one drop of $FeCl_3$ solution.
8. Observe the color of the solution.
9. Add 2mL of distilled water.
10. Observe the color of the solution again.

Observations and interpretation

The formation of a muddy or brownish violet color indicates that the sample contains hydrolyzed ester groups; yellow or no color is negative. Acrylics do not interfere but some epoxies may produce color. However, ester plasticizers in poly(vinyl chloride) will give positives. Waxed fibers or threads may give a false positive. In large samples, the final color is much closer to a dark brown with less of the violet showing up. If the solution in the end appears dark brown or black, try diluting it even further with distilled water. The diluted solution may appear more purple and thus indicate the presence of a polyester.

Storage and reagent shelf life

Hydrochloric acid (HCl) is stable when stored in a sealed, acid-resistant container in ambient conditions. Hydroxylamine hydrochloride is stable in ambient conditions, but decomposes slowly when moist. Iron(III) chloride is stable in ambient conditions, but it will react with water to produce toxic and corrosive fumes. Methanol (methyl alcohol) is stable in sealed ambient conditions. Potassium hydroxide is stable in ambient conditions. All reagent solutions are reasonably stable when stored in sealed containers at room temperature for about a month.

Test for polyamides using *p*-dimethylaminobenzaldehyde

Purpose
To identify polyamides such as nylon. This test will distinguish polyamides from other plastics such as polycarbonates, polyesters, polyethylene, and polystyrene. This test is useful in identifying any number of plastic materials including packing, storage, and display materials, and the materials used to make small articles such as jewelry, watch crystals, eyeglass lenses, buttons on textiles, etc.

Principle
When the sample is pyrolized, gases are evolved that react with *p*-dimethylaminobenzaldehyde in the presence of hydrochloric acid to form a red-colored complex on filter paper.

$$\left[-\overset{H}{N}-(CH_2)_n-\overset{H}{N}-\overset{O}{\underset{\|}{C}}-(CH_2)_n-\overset{O}{\underset{\|}{C}}-\overset{H}{N}- \right] \xrightarrow{\text{heat}} \text{gases}$$

a polyamide
(nylon)

gases + [H₃C, H₃C-N-C₆H₄-C(=O)-H] $\xrightarrow{\text{acid}}$ red complex

p-dimethylaminobenzaldehyde

Equipment

- 50 mL beaker
- droppers
- graduated cylinder
- Pasteur pipettes
- source of flame (Bunsen burner, alcohol lamp)
- qualitative filter paper
- balance (to weigh 0.1g)
- laboratory wrapping film

Reagents and safety

- **Ethanol** (ethyl alcohol): irritant and flammable; slight health rating
- **p-dimethylaminobenzaldehyde** [(*p*-(dimethylamino) benzaldehyde, 4-dimethylaminobenzenecarbanol or Ehrlich's reagent]: irritant and a mutagen; severe health rating
- **Hydrochloric acid** (HCl): toxic and corrosive; severe health rating

Protection
Wear goggles, gloves and protective clothing when handling HCl and *p*-dimethylaminobenzaldehyde. Open flames in a conservation lab are always a potential hazard. Extreme caution should be used when using any type of open flame. There should be no solvent containers in the vicinity or any other materials that can catch fire. Wear eye protection when performing this test. The fumes from the burning sample should not be inhaled. Perform test in well-ventilated area or fume hood, if possible.

Other tests to consider or confirm results

- Test for polycarbonates using *p*-dimethylaminobenzaldehyde (p. 172).
- Test for polyester groups using hydroxylamine hydrochloride (p. 168).

Reference
Rémillard, France. 1995. *La Preservation des Objects de Plastique et de Caoutchouc. Seminaire De Formation*. Quebec: Centre de Conservation du Quebec.

Reagent preparation

- **14% (w/v) *p*-dimethylaminobenzaldehyde solution**: dissolve 1.4g of *p*-dimethylaminobenzaldehyde in sufficient ethanol to make up 10mL of solution.

Method of sampling

A small sample must be removed for testing; a few fibers from textiles, a few shavings from plastic sheet, or a small piece of film (about 5mm by 5mm) should be sufficient.

Procedure

1. Seal the tip of a Pasteur pipette by melting the tip in a flame.
2. Introduce the sample into the pipette, and push it down to where the taper in the pipette begins.
3. Cut a strip of filter paper (approximately 0.6cm by 2cm), fold it slightly lengthwise, and insert it into the wide end of the pipette. The filter paper should be held in place by resistance.
4. Apply two drops of the 14% *p*-dimethylaminobenzaldehyde solution to the filter paper, followed by one drop of concentrated hydrochloric acid. Alternatively, this can be done simply by combining two drops of the 14% *p*-dimethylaminobenzaldehyde solution with one drop of concentrated hydrochloric acid, and applying this solution to the filter paper with a dropper. *Note*: The filter paper should be saturated with this solution, but not dripping down the inside of the pipette.
5. Cover the open end with a piece of laboratory wrapping film.
6. Heat the sample in the pipette, using the tip of the flame. The sample should quickly char and burn, generating smoke inside the pipette. Allow the smoke to flow over the filter paper. (*Caution*: to do this it may be necessary invert the pipette; wear gloves.)
7. Observe the color change on the filter paper.

Observations and interpretation

If the filter paper turns red the sample is a polyamide, such as nylon or a protein; if the filter paper turns blue the sample is a polycarbonate. Polyester, polyethylene, and polystyrene do not give a colored result.

Storage and reagent shelf life

The *p*-dimethylaminobenzaldehyde is light-sensitive and should be stored in the dark or in an aluminum foil-covered bottle. The 14% *p*-dimethylaminobenzaldehyde solution is stable if stored in a tightly sealed container in ambient conditions. This solution is light-sensitive and should be stored in an aluminum foil-covered bottle. Hydrochloric acid (HCl) is stable when stored in a sealed, acid-resistant container in ambient conditions.

Test for polycarbonates using *p*-dimethylaminobenzaldehyde

Purpose
To identify polycarbonates. This test will distinguish polycarbonates from other plastics such as polyamides, polyesters, polyethylene, and polystyrene. This test is useful in identifying any number of plastic materials including packing, storage, and display materials, and the materials used to make small articles such as jewelry, watch crystals, eyeglass lenses, buttons on textiles, etc.

Principle
When the sample is pyrolized, gases are evolved which react with *p*-dimethylaminobenzaldehyde in the presence of hydrochloric acid to form a blue-colored complex on filter paper.

$$[\text{aromatic group} - O - \overset{O}{\underset{\|}{C}} - O]\ \xrightarrow{heat}\ gases$$

aromatic group =

a polycarbonate

$$gases + \text{(H}_3\text{C)}_2\text{N-C}_6\text{H}_4\text{-CHO}\ \xrightarrow{acid}\ \text{blue complex}$$

Equipment

- 50mL beaker
- droppers
- graduated cylinder
- Pasteur pipettes
- source of flame (Bunsen burner, alcohol lamp)
- qualitative filter paper
- balance (to weigh 0.1 g)
- laboratory wrapping film

Reagents and safety

- **Ethanol** (ethyl alcohol): irritant and flammable; slight health rating
- ***p*-dimethylaminobenzaldehyde** [(*p*-(dimethylamino) benzaldehyde, 4-dimethylaminobenzene-carbanol, or Ehrlich's reagent]: irritant and a mutagen; severe health rating
- **Hydrochloric acid** (HCl): toxic and corrosive; severe health rating

Protection
Wear goggles, gloves and protective clothing when handling HCl and *p*-dimethylaminobenzaldehyde. Open flames in a conservation lab are always a potential hazard. Extreme caution should be used when using any type of open flame. There should be no solvent containers in the vicinity or any other materials that can catch fire. Wear eye protection when performing this test. The fumes from the burning sample should not be inhaled. Perform test in well-ventilated area or fume hood if possible. Also note that the plastic is likely to melt onto the bottom of the test tube which renders it useless for further work.

Other tests to consider or confirm results

- Test for polyamides using *p*-dimethylaminobenzaldehyde (p. 170).
- Test for polyester groups using hydroxylamine hydrochloride (p. 168).

Reference
Rémillard, France. 1995. *La Preservation des Objects de Plastique et de Caoutchouc. Seminaire De Formation*. Quebec: Centre de Conservation du Quebec.

Reagent preparation

- **14% (w/v) *p*-dimethylaminobenzaldehyde solution:** dissolve 1.4g of the solid in sufficient ethanol to make up 10mL of solution.

Sampling method

A small sample must be removed for testing. A few fibers from textiles, a few shavings from plastic sheet or a small piece of film (about 5mm by 5mm) should be sufficient.

Procedure

1. Seal the tip of a Pasteur pipette by melting the tip closed in a source of flame.
2. Introduce the sample into the pipette, and push it down to where the taper in the pipette begins.
3. Cut a strip of filter paper (approximately 0.6cm by 2cm), slightly fold it lengthwise, and insert it into the wide end of the pipette. The filter paper should be held in place by resistance.
4. Apply two drops of the 14% *p*-dimethylaminobenzaldehyde solution to the filter paper, followed by one drop of concentrated hydrochloric acid. Alternatively, this can be done simply by combining two drops of the 14% *p*-dimethylaminobenzaldehyde solution with one drop of concentrated hydrochloric acid, and applying this solution to the filter paper with a dropper. *Note*: The filter paper should be saturated with this solution, but not dripping down the inside of the pipette.
5. Cover the open end with a piece of laboratory wrapping film.
6. Heat the sample in the pipette, using the tip of the flame. The sample should quickly char and burn, generating smoke inside the pipette. Allow the smoke to flow over the filter paper (caution: to do this, it may be necessary to invert the pipette; wear gloves).
7. Observe the color change on the filter paper.

Observations and interpretation

If the filter paper turns blue the sample is a polycarbonate; if the filter paper turns red the sample is a polyamide, such as nylon or a protein. Polyester, polyethylene, and polystyrene will not give a colored result.

Storage and reagent shelf life

The *p*-dimethylaminobenzaldehyde is light-sensitive and should be stored in the dark or in an aluminum foil-covered bottle. The 14% *p*-dimethylaminobenzaldehyde solution is stable if stored in a tightly sealed container in ambient conditions. This solution is light-sensitive and should be stored in an aluminum foil-covered bottle. Hydrochloric acid (HCl) is stable when stored in a sealed, acid-resistant container in ambient conditions.

Test for silicone-based rubber using pyrolysis

Purpose

To distinguish silicone-based rubber from other rubber. Today, many silicone rubbers are pigmented black so that they look very similar to typical rubbers – it is therefore difficult to tell the two apart. Silicone-based rubbers, however, are better for use as gaskets in display cases and for any prolonged use on or near artefacts because they do not off-gas harmful chemicals. This test may help distinguish the two types of rubber. Architectural conservators may use this test to distinguish silicone caulking from other types of sealant.

Principle

Silicone rubbers contain silicon that is not combustible. Silicone rubbers will, when burned, form a residue of silicon dioxide (SiO_2) that is white.

$$\text{silicone rubber} + O_2\ (g) \xrightarrow{\text{flame}} SiO_2\ (s) + CO_2 + H_2O$$
$$\text{(white smoke and ash)}$$

Equipment

- source of flame (alcohol burner, lighter, matches)
- beaker
- tweezers

Reagents required

None.

Protection

An open flame in a laboratory can be dangerous. Precautions should be taken so that there are no containers of solvents near the flame. Goggles should be worn to protect the eyes from sparks. Rubber burns very hot and continues to burn on the inside long after the flame has been extinguished. The burned sample should be placed in the container of water to make sure that it is thoroughly extinguished before it is disposed of.

Other tests to consider or confirm results

None.

Reference

Carroll, Scott, and Leo Kohn. 1994. Silicone vs. Rubber Gaskets. *Western Association for Art Conservation Newsletter* **16**(2): 19.

Reagent preparation
None.

Method of sampling
A tiny sample of the material must be burned.

Procedure

1. Place a container of water in the area where the test is being undertaken (for extinguishing the burning sample).
2. With a pair of tweezers, hold a small sample of the material in a flame.
3. Remove it from the flame after it has started to burn.
4. Observe the characteristics of the ash.
5. Immerse the sample remains in the container of water to extinguish it. This is a safety issue as some rubber samples may continue to burn without obvious flame.

Observations and interpretation
When silicone rubber is burned, the main characteristic that distinguishes it from other rubber is the white ash that is produced. Other rubbers such as neoprene form a black sooty ash. Pure silicone rubber produces a white smoke and does not burn outside of the flame. Other rubbers give off a black sooty smoke and they continue to burn outside the flame. Some rubbers are mixtures of silicone and other rubbers and therefore show the characteristics of both.

Storage and reagent shelf life
None.

7 Other tests

pH	178
Volatile acids	182
Character	188

Test of pH using an Insta-Check® pH pencil

Purpose
To determine the pH of a substrate, solution, or vapor.

Principle
The 'lead' of this pencil contains a mixture of pH indicators in a water-soluble wax. When the pencil 'lead' comes into contact with water, the indicators go into solution and react with the material being tested.

Indicators for pH are usually complex organic molecules that behave as acids or bases. When exposed to an acid, for example, the indicator accepts a proton (acting like a base). The protonated form of the indicator absorbs light of a different wavelength changing the color of the solution or substrate. The opposite reaction takes place when the indicator is exposed to a base. The indicator donates a proton (acting like an acid) and the new species is yet a different color. Each indicator molecule gains or loses a proton at a very specific pH range, therefore a range of pH can be determined by using a mixture of these indicator molecules.

$$\text{substrate (acidic or basic)} + \text{indicator (pencil)} = \text{color change}$$

$$\text{acid} + \text{indicator} \rightarrow [\text{H}+\text{indicator}]^+$$

(red to yellow)

$$\text{base} + \text{indicator} \rightarrow [\text{indicator--H}]^-$$

(green to violet)

Equipment

- hot plate
- beaker
- sealed container
- filter paper

Reagents

- **Insta-Check® pH pencil** [Microessential Laboratory, distributed by University Products, Talas or Preservation Equipment, UK]
- **Water** [distilled, boiled]

Other tests to consider or confirm results

- Test for acidity with pH pens (p. 180).

Reference
Mayer, Debora D. (compiler) 1990. Spot Tests. In *Paper Conservation Catalog*, 7th edition, 21–2. Washington DC: American Institute for Conservation Book and Paper Group.

Reagent preparation

- **Boiled distilled water (BD):** distilled water should be boiled for at least 5 minutes to remove most of the carbon dioxide (CO_2). This important step should not be ignored. When CO_2 dissolves in water it forms carbonic acid ($CO_2 + H_2O \rightarrow H_2CO_3$) which will lower the pH of the water. The pH of distilled water (when purchased from the store) is usually slightly acidic (~pH6) because it has absorbed CO_2 from the air. By boiling the water, the pH increases and sometimes it will rise to pH7. The water should be allowed to cool in a container with a tight lid to prevent more CO_2 from dissolving in it while cooling. De-ionized water may be used if distilled water is not available but it must also be boiled first.
- **pH indicator solution**: the pH pencil can be used as is or a solution can be made from the pencil 'lead'. Place a few shavings in the bottom of a test tube, add a few drops of BD water – the 'lead' dissolves readily. The advantage of this method is that the pH of solutions may be tested. Also the pencil can be stored or carried into the field and then made into solution when needed.

Method of sampling
The test is performed directly on the surface of the object.

Procedure
For checking paper type products
1. Make a small dot on the surface of the material with the pencil.
2. Place a drop of BD water on it.

For checking pH of solutions
1. Place 2mL of the sample solution in a clean test tube that has been rinsed with BD water.
2. Add one drop of the pH indicator solution.
3. Agitate the solution in the test tube.

For checking acid vapors
1. Dip a piece of filter paper in the pH solution and allow paper to dry.
2. Place test paper near to but not touching object(s) to be tested.
3. Observe color change and compare with color strip.

Observations and interpretation
The colors obtained in the test should be compared with the color strip provided with the pH pencil. In general, the colors should indicate as the following pH: red-orange 0–3; yellow 4–6; yellow-green 7; green 8; dark green 9; dark green blue 10; dark blue 11–12; violet 13. If there are standard pH solutions available (pH4, 7 and 10) a few milliliters of each can be placed in separate test tubes with a drop of pH indicator solution and compared as reference solutions.

Storage and reagent shelf life
The pencil is stable. The solution made up from the pencil can be kept in a sealed container at room temperature and out of direct light for a few days.

Test for acidity with pH pens

Purpose
To determine if paper, interleaving tissue, or paper board is safe for use in archival storage.

Principle
The 'ink' of proprietary pH pens contains a pH indicator (usually chlorophenol red or bromophenol green) which changes color at specific pH. pH indicators are usually complex organic molecules that can behave as acids or bases. When exposed to an acid, for example, they accept a proton (acting like a base). The protonated form of the indicator absorbs light of a different wavelength changing the color of the substrate. The opposite reaction takes place when exposed to a base. They donate a proton (acting like an acid) and the new species is yet a different color. Each indicator molecule gains or loses a proton at a very specific pH.

$$acid + indicator \rightarrow [H+indicator]^+$$
$$(red\ to\ yellow)$$

$$base + indicator \rightarrow [indicator-H]^-$$
$$(green\ to\ violet)$$

Equipment
None.

Reagents

- **Abbey pH pen** [University Products, Holyoke, MA, cat no 3-0019 or Preservation Equipment, UK]
- **pH Pen** [Light Impressions, Rochester, NY or Preservation Equipment, UK]

Other tests to consider or confirm results

- Test of pH using an Insta-Check® pH pencil (p. 180).

Reference
Mayer, Debora D. (compiler) 1990. Spot Tests. In *Paper Conservation Catalog*, 7th edition, 21–22. Washington DC: American Institute for Conservation Book and Paper Group.

Reagent preparation
None.

Method of sampling
The test may be carried out directly on the surface of the material to be tested or the test sample may be torn to expose the inside surface for testing.

Procedure

1. Make a line on the sample with the pen and observe the color as soon as it dries (30 seconds). Make sure that the ink has been absorbed by the paper, sometimes two or three strokes are necessary.

Observations and interpretation
Follow the instructions that come with the pen. Pens from different manufacturers form different colors in the presence of acids. The test works best in the determination of whether a paper contains a buffer or not. The test does not work well as a true indicator of pH since there really are only two colors which indicate acidic or basic conditions. Obviously if the sample is colored the indicator may not be visible or a slightly different color may be produced. Sometimes, when a visually undetectable treatment or coating has been applied to the surface of paper, the surface has a pH that is much higher or lower than the paper below. In this case, a more accurate result is achieved if the inside surface it tested.

Storage and reagent shelf life
Stable.

Test for volatile acids with pH papers and lime water

Purpose
To determine the presence of volatile acids in materials being considered for storing or exhibiting museum collections.

Principle
The release of acids from the sample will neutralize the base, calcium hydroxide ($Ca(OH)_2$), by forming a salt. This will change the color of the pH paper to give a more acidic reading.

$$acid + base \rightarrow salt + H_2O$$

$$2HA\ (aq) + Ca(OH)_2\ (aq) \rightarrow CaA_2\ (aq) + 2H_2O\ (l)$$
$$\text{acid} \qquad \text{base} \qquad \text{salt}$$

Indicators for pH are usually complex organic molecules that can behave as acids or bases. When exposed to an acid, for example, the indicator accepts a proton (acting like a base). The protonated form of the indicator absorbs light of a different wavelength changing the color of the substrate.

$$acid + indicator \rightarrow [H+indicator]^+$$
$$\text{(red to yellow)}$$

It is important to run a control for this test and to use the color on the pH paper of the control to compare with the color developed on the pH paper in the sample test tube. Time is also an important factor in this test. The pH indicated on the control will also decrease with time because of CO_2 in the air reacting with the pH paper.

Equipment

- 50mL test tubes with screw caps (polyethylene stoppers or laboratory sealing film will also do for sealing the tubes)
- pH indicator paper with the range from pH0–14
- polyester batting
- graduated cylinder
- balance (to weigh 0.1g)
- oven or warm area
- cotton swabs
- reagent bottle
- hot plate
- beaker

Reagents and safety

- **Calcium hydroxide ($Ca(OH)_2$):** corrosive and an irritant; slight health rating
- **Water** [distilled, boiled]

Other tests to consider or confirm results

- Tests for volatile acids with pH papers and glycerol (p. 184).

Reference
Hopwood, Walter R. 1979. Choosing Materials for Prolonged Proximity to Museum Objects. *American Institute for Conservation 7th annual meeting*, Toronto, 44–9. Washington, DC: American Institute for Conservation, pre-print.

Reagent preparation

- **Boiled distilled water (BD water):** distilled water should be boiled for at least five minutes to remove all carbon dioxide (CO_2) This important step should not be ignored. When CO_2 dissolves in water it forms carbonic acid ($CO_2 + H_2O \rightarrow H_2CO_3$) which will lower the pH of the water. The pH of distilled water (when purchased from the store) is usually slightly acidic (~pH6) because it has had a chance to absorb CO_2 from the air. By boiling the water the pH increases, sometimes it will rise to 7. The water should be allowed to cool in a container with a tight lid to prevent more CO_2 from dissolving in it while cooling. De-ionized water can be used if distilled water is not available but it also must be boiled first.
- **Calcium hydroxide solution (lime water):** dissolve 0.5g calcium hydroxide ($Ca(OH)_2$) in 100mL of BD water.

Method of sampling

Some of the material must be removed for testing.

Procedure

1. Add 1mL BD water to each of two test tubes (one is the sample and one is the control).
2. Place a small piece of polyester batting in each test tube to the 10mL level. This prevents the sample from getting wet.
3. Add 1g of material to be tested to one of the test tubes.
4. Use a cotton swab dipped in the lime water to wet the pH paper (it should read pH10–12) for each test tube.
5. Insert the pH indicator paper into the test tubes so that it stays in the upper section of the test tube. The stiff plastic pH strip has enough resistance when folded over to hold itself in place against the walls of the test tube. It is preferable to fold it so that the indicator portion is on the inside where it will have the best chance of contact with the atmosphere. Another way is to attach the pH indicator paper to a smaller vial with Teflon tape. The smaller vial should fit easily inside the larger test tube, the excess Teflon tape holding the smaller vial in place inside the test tube.
6. Seal the test tubes with stopper, cap or laboratory sealing film.
7. Place the test tubes in a 50°C oven or warm area.
8. Read and record pH from paper strips immediately, after 1 hour, 4 hours, 1 day and several days.

Observations and interpretation

Decreasing pH is evidence of volatile acids being released from the material being tested. The test should be closely monitored every hour for the first four hours and compared with the control to see if there is a difference that is attributable to the material.

Storage and reagent shelf life

Lime water is stable when stored in a sealed container in ambient conditions for short periods. BD water should be freshly prepared before each test.

Test for volatile acids with pH papers and glycerol

Purpose
To determine the presence of volatile acids in materials being considered for storing or exhibiting museum collections.

Principle
The presence of a volatile acid will change the color of pH paper by reacting with indicator compounds on the paper.

$$\text{acid} + \text{indicator} \rightarrow [\text{H}+\text{indicator}]^+$$

$$\text{(red to yellow)}$$

Indicators for pH are usually complex organic molecules that can behave as acids or bases. When exposed to an acid, for example, the indicator accepts a proton (acting like a base). The protonated form of the indicator absorbs light of a different wavelength changing the color of the substrate. The opposite reaction takes place when an indicator molecule is exposed to a base. The indicator donates a proton (acting like an acid) and the new species is yet a different color. Each indicator molecule gains or loses a proton at a very specific pH range. Therefore a range of pH can be determined by using a mixture of these indicator molecules.

Treating the paper with glycerol/water solution increases the adsorption of acids from the surrounding atmosphere and makes it more sensitive and faster to register the presence of volatile acids. If only distilled water is applied to the indicator paper, the water will evaporate after a while and the indicator paper will no longer function. If pure glycerol is used on the pH indicator, a glycerin water droplet will form on the surface of the indicator paper as the hygroscopic glycerin absorbs water vapor. This droplet could distort the results.

Note: chemically glycerol and glycerin describe the same compound and can be used interchangeably.

Equipment

- 50mL test tubes or vials that can be sealed
- laboratory wrapping film or polyethylene stoppers
- glass stirring rod
- pH indicator paper with a 0–14 range
- balance (to weigh 1.g)
- cotton swabs
- reagent container
- graduated cylinder
- hot plate
- beaker

Reagents and safety

- **Glycerol** (glycerin): slight health rating
- **Water** [distilled, boiled]

Other tests to consider or confirm results

- Test for volatile acids with pH papers and lime water (p. 182).

Reference
Tetreault, Jean. 1992. La Mesure de L'acidite des produits volatils (Measuring the Acidity of Volatile Products). *Journal of International Institute for Conservation-Canadian Group* 17: 17–25.

Reagent preparation

- **Glycerol solution**: mix 80mL glycerol and 20mL boiled distilled water (BD water).
- **BD water**: distilled water should be boiled for at least 5 minutes to remove all carbon dioxide (CO_2). See page 183.

Method of sampling
Some of the material must be removed for testing. If testing a large space (such as a storage cabinet), leave the pH strip suspended inside.

Procedure

1. Place 1g of the material in a test tube or vial and leave another empty as a control.
2. Apply a very thin layer of the glycerol solution to two pH indicator papers with a cotton swab.
3. Insert the pH indicator papers into the test tubes so that they stay in the upper section of the test tube. The stiff plastic pH strip has enough resistance when folded over to hold itself in place against the walls of the test tube. It is preferable to fold it so that the indicator portion is on the inside where it will have the best chance of contact with the atmosphere. Another way is to attach the pH indicator paper to a smaller vial with Teflon tape. The smaller vial should fit easily inside the larger test tube, the excess Teflon tape holding the smaller vial in place inside the test tube.
4. Seal the test tubes with stopper, cap or laboratory wrapping film.
5. Monitor every day.

Observations and interpretation
If volatile acids are released by the material in question or are present in the cabinet, the pH indicated on the paper will be lower than that indicated on the blank test. After a few days a difference in the two papers should be noted. For this test it is important to have a control for comparison. The control should not change from the initial reading (most pH papers that indicate in the 0–14 range initially exhibit a pH somewhere around pH 5). It is useful to test a material that is known to give off volatile acids (fresh pine wood shavings for example) for comparison. For certain materials the test may take a few weeks. It is also important to note that this test is only qualitative (not quantitative). This means that it will indicate the presence of volatile acids but will not give the definite pH of the material being tested.

Storage and reagent shelf life
All materials used are stable, however some pH papers are light-sensitive and should not be kept in direct light for long periods of time.

Test for acidic vapors using cresol red

Purpose
To determine if objects (such as cellulose nitrate) are evolving acidic vapors so that precautions may be taken to protect the object itself or others in the collections from the vapors.

Principle
Strongly acidic vapors react with certain indicator dyes to change their color. Paper impregnated with these dyes may be placed close to a suspect object and observed for a change of color in the paper. The test is recommended particularly for detecting NO_2 from degrading cellulose nitrate.

$$\text{cellulose nitrate} \xrightarrow{\text{time}} NO_2 \ (g)$$

$$2NO_2 \ (g) + H_2O \ (g \text{ or } l) \rightarrow HNO_3 \ (aq) + HNO_2 \ (aq)$$
$$\text{nitric acid} \quad \text{nitrous acid}$$

$$\text{acid} + \text{indicator} \rightarrow [H + \text{indicator}]^+$$
$$\text{cresol red} \quad \text{(yellow to pink)}$$

Equipment
- filter paper or cotton string
- reagent container
- glass stirring rod
- balance (to weigh 0.001g)

Reagents and safety
- **Cresol red or cresol purple**: irritants; slight health rating
- **Ethanol** (ethyl alcohol): flammable and an irritant; slight health rating
- **Methanol** (methyl alcohol): toxic, flammable and irritant; severe health rating

Protection
Wear goggles, gloves and protective clothing when handling methanol.

Other tests to consider or confirm results
- Test for nitrate using spot test papers (p. 112)
- Test for nitrate (cellulose nitrate) using diphenylamine (p. 164).

Reference
Fenn, Julia. 1996. 'The Cellulose Nitrate Time Bomb: Using Sulphonephthalein Indicators to Evaluate Storage Strategies' in *From Marble to Chocolate: The Conservation of Modern Sculpture*, Jackie Heuman (editor) 87–92. London: Archetype.

Reagent preparation

- **0.005% cresol red solution**: add 0.005g to a mixture of 10mL methanol and 90mL ethanol.

Method of sampling

This test does not require that a sample be removed from the object. The impregnated test paper is placed near to but not touching the material being tested.

Procedure

1. Dip the paper or string in the dye solution and remove excess by tapping with a glass rod. Allow the impregnated test paper to dry completely.
2. Place the test paper or suspend the string so that they are near to but not touching the object(s) to be tested.
3. Observe for color change.

Observations and interpretation

The dye in the paper will change color if acidic vapors are present. The color change will be more intense near areas that emit more acidic vapors. Generally the color change will occur within one or two days, but some samples may take up to five days to produce a color reaction. It is best if the paper is placed beneath the material being tested as acidic vapors are heavier than air and tend to sink. The paper will change from yellow to reddish pink. Fresh cellulose nitrate (non-degraded) does not give a positive result with this test. A strong basic vapor will change the paper from yellow to violet.

Storage and reagent shelf life

Cresol red and purple solutions are stable when stored in sealed containers in ambient conditions.

Test for hardness with a pencil sequence

Purpose
This test is best used to characterize hardness and compare it to knowns that are tested at the same time. It will also be useful to help determine if two materials are the same or not (such as wood, bone, horn, plastic films and adhesives, paints, and coatings, etc.).

Principle
Pencils of varying hardness are used to try to scratch or penetrate the surface of an object; the hardness is represented by the hardest pencil that will not scratch the object or penetrate it. The sequence from softest to hardest is as follows: 6B, 5B, 4B, 3B, 2B, B, HB, F, H, 2H, 3H, 4H, 5H, 6H.

Equipment
- Many suppliers make graded pencils, but for this test the following types are recommended: Microtomic by Faber Castell or Turquoise-T2375 made by Mohawk
- vinyl eraser

Reagents
None.

Other tests to consider or confirm results
None.

Reference
American Society for Testing and Materials. 1989. Standard Test Method for Film Hardness by Pencil, Test # D 3363-74, reapproved. *Annual Book of Standards*, vol 6.01, 518–19. West Conshohocken, PA: American Society for Testing and Materials.

Reagent preparation
None.

Method of sampling
The test is performed directly on the surface of the object.

Procedure
1. Start with the softest pencil (6B). The leads should not come to a sharp point, but should be slightly flat.
2. In an inconspicuous place rub the pencil lead back and forth in one area.
3. Use the eraser to wipe off the pencil mark. Examine the surface to see if the pencil left a scratch.
4. Use successively harder pencils until there is one that leaves a scratch on the surface after the mark has been erased.

Observations and interpretation
Report hardness in terms of the hardest pencil that makes no permanent mark. Thus if 2H left a permanent scratch, the hardness of the material would be H. The sequence from softest to hardest is as follows: 6B, 5B, 4B, 3B, 2B, B, HB, F, H, 2H, 3H, 4H, 5H, 6H.

Storage
The pencils will store indefinitely.

Test to determine specific gravity using an electronic analytical balance

Purpose
To determine the specific gravity of an object using a platform balance. Specific gravity can aid in characterization of materials such as plastics, metals and minerals.

Principle
An object is weighed (or a small sample if the object is very large) in air and then again in a liquid of known specific gravity (SG_k). The object must be more dense than the liquid but should not swell or dissolve in it. The sample must be non-porous, solid, without air pockets or bubbles. If the sample contains holes or crevices then care must be taken to make sure that the liquid penetrates all of them. The specific gravity of the unknown sample (SG_u) can then be calculated from the following equation:

$$SG_u = \left(\frac{\text{mass of substance in air}}{\text{loss of mass in liquid}} \times SG_k \right) \text{ at room temperature}$$

SG_u = specific gravity of unknown, SG_k = specific gravity of known liquid

Loss of mass in liquid represents the difference in weight between the mass of the substance in air and the mass of the liquid of known specific gravity, (weight_{air} – weight_{liquid}).

Some examples of specific gravity values

White microcrystalline wax	0.928–0.941	Vegetable ivory	1.38–1.42
Natural rubber	0.93	Cellulose nitrate (nitrocellulose)	1.58–1.66
Linseed oil (*Linum usitatissimum*)	0.938	Elephant ivory	1.70–1.98
White beeswax USP	0.959–0.975	Bone	1.94–2.10
Singapore damar	1.04–1.06	Natural pearls	below 2.173
Polystyrene	1.04–1.07	Cultured pearls	above 2.173
Amber	1.08	Coral	2.60–2.70
Poly(vinyl acetate)	1.17–1.19	Aluminum	2.70
Poly(vinyl alcohol)	1.21–1.32	Hematite	5.26
Tortoise shell	1.29	Zinc	7.14
Epoxy casting resins	1.3–1.4	Nickel	8.89
Jet or black amber	1.33	Lead	11.35
Poly(vinyl chloride)	1.38–1.41	Gold	19.3

Equipment

- balance (to weigh 0.01g)
- thin wire
- hook
- foam board
- rigid polyethylene foam
- beaker

Reagents and safety

- **Water** [de-ionized]
- Or a liquid of known specific gravity, such as ethylene dibromide

(Safety precautions: none if water is used; if organic solvents are used then appropriate precautions should be taken.)

Other tests to consider or confirm results
None.

Reference
Moss, A. A. 1956. *The Identification of Metals. Handbook for Museum Curators*, Part B, *Museum Technique*, Section 8, 2–8. London: Museums Association.

Preparation

If a specific gravity attachment is not available, it will be necessary to first construct a device that allows the object to be weighed in water (see Fig. 7, p. 27). This can best be made from foam board and Styrofoam to reduce its weight.

- Cut a disk about the size of the balance platform from foam board to serve as a base for the armature.
- Cut a 2cm × 2cm × 10cm post from rigid polyethylene foam and attach the post to the centre of the base.
- Place a foam board arm (parallel to the platform) on the top of the post and add a small hook to the end on the underside.
- Cut a crescent-shaped piece of foam board and attach three legs of rigid polyethylene foam about 1cm × 2cm to one side. The legs should be large enough to allow the crescent to stand above the balance pan and the base of the armature. This will support a small beaker.
- Use thin wire to suspend a small basket (made of aluminum foil with small perforations in the centre) from the hook. It is important that the basket does not touch the beaker.

Method of sampling

Sampling is not necessary if the object is small, like a bead. Otherwise a sample must be removed from the object. Objects with SG_u less than 1 will float in water and a lower density liquid will be necessary for the procedure.

Procedure

1. Place the specific gravity device and an empty beaker in the weighing chamber. The armature rests on the weighing pan of the balance and the crescent-shaped platform rests around but not touching the weighing pan. Place a beaker on this platform.
2. Weigh the armature and empty basket (W_1).
3. Place the object or sample into the basket and weigh (W_2). The weight of the sample in air, $W_3 = W_2 - W_1$.
4. Fill the beaker with de-ionized water so that the sample is immersed in it. The sample or the basket should not touch the beaker (wait about a minute to let any air bubbles escape).
5. Record the weight of the object while immersed in the liquid (in water) (W_4). Caution: air trapped in a sample can affect the accuracy of the weight measurement.
6. Remove the sample and re-weigh and record the weight of the basket immersed in water (W_5).
7. Subtract the weight of the basket in water (W_5) from the weight of the sample in the basket in water (W_4) to get the weight of the sample in water ($W_6 = W_4 - W_5$).
8. Then subtract the weight of the sample in water (W_6) from the weight of the sample in air (W_3). The loss in mass, $W_7 = W_3 - W_6$.
9. Repeat the above procedures several times and compare the results.

Observations and interpretation

Calculate the specific gravity:

$$SG_u = \left(\frac{\text{mass of object in air (step 3 or } W_3\text{)}}{\text{loss of mass in liquid (step 8 or } W_7\text{)}} \times SG_k\right) \text{ at the temperature of the measurement}$$

To reduce the surface tension and facilitate wetting of the object a very dilute solution (<0.01%) of a surfactant is desirable. It can be assumed that this does not change the density of water (which for this test, can be considered to be 1.00 at room temperature).

Test for radioactivity using photographic film

Purpose
To detect radioactivity in a specimen such as a mineral or glass that contains uranium, or other radioactive materials.

Principle
Radioactivity is the result of natural or artificial disintegration of nuclear material by the emission of radiation in the form of alpha (α), beta (β) or gamma (γ) rays. These will affect photographic film. Exposure of the film shows the presence of radioactivity.

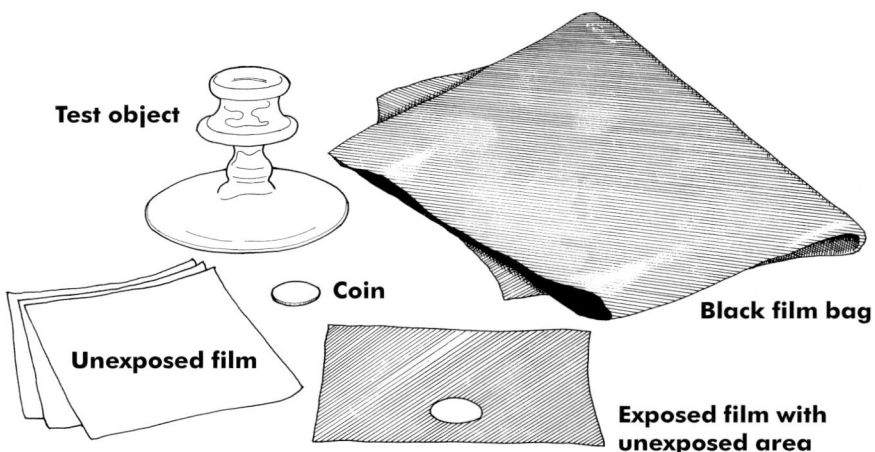

Equipment

- light-protecting container (film canister, film bag or lead bag used to protect photographic film from airport X-ray machines)
- photographic film processing (equipment used for film development if developing in-house)
- paperclips or coins
- tissue paper

Reagents and safety

- **Black and white photographic film** (T-MAX 400 ISO speed 4" × 5" black and white sheet works well)
- Standard chemistry used for film development if developing in-house

Radioactivity can be dangerous; if samples exhibit activity follow standard OSHA or safety procedures for storage and handling.

Other tests to consider or confirm results
None.

Reference
Blount, Alice. M. 1990. A Low-cost Radioactivity Test for Geological Specimens. *Collection Forum* 6(1): 8–11.

Reagent preparation
None.

Method of sampling
The entire object is tested in a non-destructive fashion.

Procedure
For small samples
1. Cut off 3–4 inches of black and white film from a roll in a darkroom.
2. Place a paperclip on it and roll, and place in a 35mm film canister.
3. Wrap the sample in some tissue paper and insert into canister.
4. Close tightly before exposing to light.
5. After three to four days, remove film in the dark, and develop film according to the manufacturer's instructions.

For larger samples
1. In the dark, place a few paperclips or a coin on the edge of a 4" × 5" or 5" × 7" sheet of black and white film.
2. Slip the film into a lead bag (the kind used to protect photographic film from airport X-ray machines).
3. Insert the specimen (wrapped in tissue paper) and seal tightly.
4. After three to four days, remove film in the dark, and develop film according to the manufacturer's instructions.

Alternative procedure
1. Place the paperclips on three pieces of black and white photographic film.
2. Place the specimen (wrapped in tissue paper) on the film in the light-protecting container.
3. Carefully remove single sheets of film from the container after one day, three days, and six days.
4. Develop the film according to the manufacturer's instructions.

Observations and interpretation
Look for fogging (darkening) and the image of the paperclip (light shadow against darkened background). If that can be seen, the sample is radioactive. The benefit of using three pieces of film is that specimens having low-level radioactivity may require more time to visually expose the photographic film.

Storage and reagent shelf life
If developing in-house follow the manufacturer's instructions.

Appendix 1: Dilution table and chemical concentration calculations

Material characterization methods utilizing chemical spot-test techniques require the preparation of reagent solutions in varying concentrations. In the course of working with acids, bases, or salts in aqueous solution it is generally necessary to achieve a specific concentration, often by diluting a more concentrated solution. Unfortunately, in the spot-testing literature, the chemical terminology for calculating solution concentrations is not standardized and the spot-test literature usually contains diverse methods of reagent preparation.

Solutions are made when a solute (usually a solid) is dissolved in a solvent (usually a liquid). The concentration of the solution indicates the quantity of solute in the solvent. Solutions are often referred to as percent solutions and may be made in three forms: (1) volume/volume (v/v) percent, (2) weight/weight (w/w) percent, and (3) weight/volume (w/v) percent. Making accurate percent solutions for solvent mixtures is difficult because it requires a calculation of the density for each substance used in the solution. Density refers to mass (grams) of a substance divided by the volume (milliliters or cubic centimeters) it occupies.

For chemical solutions, the preferred method of describing the concentration, or [C], is in moles of solute per liter of solution. This is commonly known as the **molarity** of the solution, and is designated by the symbol M. Therefore, a 2.0M solution of sodium hydroxide (NaOH) corresponds to 2.0 moles of NaOH dissolved in enough water to make 1.0L of solution. But what does 2.0 moles of NaOH mean?

The concept of the mole is relatively simple: it is the bridge between the atomic scale (number of atoms or molecules) and the macroscopic scale (mass of a substance). It takes 6.02×10^{23} atoms or molecules to make a mole of that atom or molecule. *One mole of anything always contains the same number of particles, no matter what the substance.* This number (6.02×10^{23}), referred to as Avogadro's number, is named after the eighteenth-century Italian lawyer and physicist Amedeo Avogadro. It is significant because it relates the number of particles to their mass (grams). For example, if we examine the periodic table of elements, we find that the sodium atom (Na) has a mass number of 23.0. This mass number represents the mass (grams) of one mole of sodium atoms (or 6.02×10^{23} atoms) therefore the mass number for sodium is 23.0 grams per mole (g/mol). The same holds true for all of the atoms on the periodic table – the mass number represents the mass (grams) of one mole (6.02×10^{23} atoms) of the element (e.g. carbon is 12.0g/mol, oxygen is 16.0g/mol, etc). Often the term 'molecular weight' is used to describe this mass number. But how does this work for molecules?

Since molecules are composed of more than one atom, the mass numbers of the individual atoms can be added together to determine the molecular weight of the molecule. For example, one molecule of NaOH is composed of one hydrogen atom with a mass number of 1.0g/mol, and one oxygen atom with a mass number of 16.0g/mol, and one sodium atom with a mass number of 23.0g/mol; so the molecular weight of NaOH is $1.0 + 16.0 + 23.0 = 40.0$g/mol. Again, this molecular weight describes the mass of one mole of NaOH or 6.02×10^{23} molecules of NaOH. If we re-examine the 2.0M solution of NaOH, we find that the solution requires 2.0 moles of NaOH or 80.0 grams of NaOH.

In preparing solutions of different molarity the following formula is useful:

$$G = L \times M \times MW$$

where G = grams of solute required (g), L = liters of solution (L), M = molarity of the solution (mol/L), and MW = molecular weight of the solute. The formula may also be used to prepare solution dilutions in small quantities. For example, to prepare 100mL of a 0.20M NaOH solution we use the above equation and simply plug-in the known quantities:

$$G = (0.100L) \times (0.20M) \times (40.0) \text{ g/mol}$$

$$G = 0.8 \text{ grams of NaOH}$$

Note that we have to convert 100mL to liters and determine the molecular weight of NaOH. Therefore, to make 100mL of a 0.20M NaOH solution we must:

- weigh 0.8g of NaOH
- place this quantity in a 100mL volumetric container
- add enough water to dissolve the solid, then add additional water to the 100mL mark
- place in a tightly sealed container with a label that includes the date

It is recommended that the acid and base reagents used in chemical spot testing be prepared in small quantities. Therefore, it is convenient to consider that if a solution concentration is XM, then each milliliter (mL) or cubic centimeter (cc) will contain X millimoles (mmol). The solution concentration can also be described as X mmol/mL. This conversion is nothing more than dividing the unit of molarity (mol/L) by 1000 to give the new unit millimolarity (mmol/mL).

To illustrate these points let us prepare a 2.0M solution of hydrochloric acid (HCl) from the commercially available solution. This will require a dilution (a solution derived from a liquid solute dissolved in a liquid solvent). Commercial solutions of HCl are highly concentrated, typically between 11M and 12M. If the commercially available HCl is 11.6M, we must dilute it so that the final concentration is 2.0M. As described earlier, if a solution is X M then it is also X mmol/mL. It is more convenient to discuss our dilution in these terms. The dilution calculation determines how much water is needed to dilute 1.0mL of 11.6M to 2.0M. The dilution calculation is a simple ratio:

$$\text{dilution factor} = \frac{\text{concentration of standard solution}}{\text{desired concentration}}$$
$$= \frac{11.6\text{M}}{2.0\text{M}} = 5.8$$

This dilution factor (5.8) represents the final volume of the diluted solution if 1.0mL of concentrated HCl is used. In other words, if 1.0mL of 11.6M HCl is diluted to a total volume of 5.8mL, the final concentration will be 2.0M. This means that we need to add 1.0mL of concentrated HCl to 4.8mL of water to obtain 5.8mL of 2.0M HCl solution. Remember: **ALWAYS ADD ACID TO WATER!** But what if we need a specific volume of a diluted solution, how is this calculation done?

For example, we need to prepare 100mL of 2.0M HCl from concentrated hydrochloric acid. Another dilution calculation is necessary. We begin by recognizing that if 1.0mL of the 2.0M HCl contains 2.0mmol of HCl, then 100mL contains 200mmol of the HCl. The dilution calculation determines the volume (milliliters) of the 11.6M HCl needed to give 200mmol. This means that we must take 200mmol of concentrated HCl and dilute it to 100mL. But how do we determine the volume of concentrated HCl that contains 200mmol? The equation is:

$$\text{volume of concentrated solution (mL)} = \frac{\text{amount of solute required (mmol)}}{\text{concentration of standard solution (M)}}$$

So to make 100mL of a 2.0M HCl solution from concentrated HCl we must:

- determine the amount of solute (mmol) from the desired concentration and volume

    amount of solute (mmol) = concentration (M) × volume (mL)

    amount of solute (mmol) = (2.0M) × (100mL) = 200mmol

- divide the amount of solute (200mmol) by the concentration of the standard solution (11.6M)
- the resulting volume (17.2mL) represents the amount of concentrated HCl that must be diluted to 100mL
- fill a 100mL volumetric flask approximately halfway with water
- add the 17.2mL of concentrated HCl to the water in the 100mL flask. **ALWAYS ADD ACID TO WATER!**
- add more water to the 100mL mark
- place in a tightly sealed container with a label that includes the date

If the target concentration is much lower than the starting concentration it may be necessary to prepare one or even two intermediate concentrations. For example, if you wish to prepare a 0.033M solution from an 18M solution it would require diluting 1mL to 545.5mL. If that much reagent is not necessary, then a dilution to 1.8M (1mL diluted to 10mL) followed by a dilution of 1mL of that solution to 54.5mL will prepare the desired 0.033M solution.

In this book the dilution concentrations are written in the test protocols as ratios using two numbers and a colon. For example 1:1 is meant to mean one part concentrated reagent added to one part solvent.

Dilution table

Reagent	Dilution, v/v (reagent/H$_2$O)	Molarity (M)	Normality (N)	Concentration (drops/mL)	Specific gravity
Glacial acetic acid	100% undiluted	17.40	17.40	63	1.05
	1:1	8.70	8.70		
	1:3	4.35	4.35		
	1:7	2.18	2.18		
	1:16.4	1.00	1.00		
Hydrochloric acid	37% undiluted	11.60	11.60	23–24	1.18
	1:1	5.80	5.80		
	1:3	2.90	2.90		
	1:5	1.93	1.93		
	1:10.6	1.00	1.00		
	1:22.2	0.50	0.50		
	1:115	0.10	0.10		
Nitric acid	70% undiluted	15.40	15.40	36–37	1.42
	1:1	7.70	7.70		
	1:4	3.08	3.08		
	1:7	1.93	1.93		
	1:14.4	1.00	1.00		
	1:30	0.50	0.50		
	1:76	0.20	0.20		
Phosphoric acid	85% undiluted	14.60	43.80	25–26	1.70
	1:1	7.30	21.90		
	1:2	4.87	14.60		
	1:4	2.92	8.76		
	1:7	1.83	5.48		
	1:11	1.22	3.65		
	1:13.6	1.00	3.00		
	1:43	0.33	1.00		
Sulfuric acid	98% undiluted	17.80	35.60	36–37	1.84
	1:1	8.90	17.80		
	1:2	5.93	11.87		
	1:4	3.56	7.12		
	1:8	1.98	3.96		
	1:17	0.99	1.98		
	1:34	0.51	1.02		
Ammonium hydroxide	29% undiluted	14.80	14.80	24–25	0.89
	1:1	7.40	7.40		
	1:2	4.93	4.93		
	1:4	2.96	2.96		
	1:7	1.85	1.85		
	1:14	0.99	0.99		
	1:47	0.10	0.10		

	g/100mL	Molarity (M)	Normality (N)	Molecular weight
Sodium hydroxide	20.00	5.00	5.00	40
	10.00	2.50	2.50	
	4.00	1.00	1.00	
	0.40	0.10	0.10	

Appendix 2: Explanation of pH

Acids and bases are two classes of related chemicals and members within each class have a number of similar properties when dissolved in a solvent, which is usually water. In aqueous solutions the concentration of acids or bases can vary by many orders of magnitude by powers of 10. Chemists utilize a convenient shorthand for expressing these concentrations without resorting to the cumbersome use of strings of zeros. The Danish biochemist S. P. L. Sorensen introduced the system of representing the hydrogen ion exponent commonly known as pH. The pH of a solution is defined as the negative logarithm of the hydrogen ion concentration

$$\text{pH} = -\log [H^+]$$

The term $[H^+]$ (or the enclosure of a chemical symbol within square brackets) denotes that the concentration of the chemical species is the quantity being considered; in this case it is the hydrogen ion.

There are many complementary theories relating to acids and bases. One of the earliest concepts was developed by the Swedish chemist Svante Arrhenius in the late nineteenth century. The Arrhenius theory defines an acid as a compound that can dissociate in water to yield hydrogen ions (H^+) and a base as a compound that can dissociate in water to yield hydroxide ions (OH^-). Moreover, the degree to which an acid or base dissociates in water can be described. Strong acids and bases tend to completely dissociate in water, while weak acids and bases only minimally dissociate in water. Hydrochloric acid (HCl) and sodium hydroxide (NaOH) are examples of a strong acid and base respectively, and acetic acid (vinegar) and ammonium hydroxide (ammonia) are examples of a weak acid and base. It should also be noted that the term hydrogen ion (H^+) is misleading. In aqueous solutions a more accurate description of the acidic species is the hydronium ion (H_3O^+).

In any aqueous system some of the water molecules separate into hydronium and hydroxide ions:

$$2\ H_2O \rightarrow H_3O + OH^-$$

The concentrations of these ions are extremely small in pure water; 0.0000001M (1×10^{-7}M). We can also describe this concentration in terms of pH by taking the negative logarithm of the hydronium ion concentration

$$pH = -\log[H_3O^+] = -\log(1 \times 10^{-7}) = 7$$

This basically means that pure water is primarily made up of H_2O molecules and has a neutral pH of 7.0. But as acidic substances are added to water the hydronium ion concentration increases rapidly, while the hydroxide ion concentration is effectively reduced:

$$acid + H_2O \rightarrow [\mathbf{H_3O^+}]\uparrow + [OH^-]\downarrow$$

Let us say that the new hydronium ion concentration has increased to 0.01M (1×10^{-2}M). Again, this can be written as a pH:

$$pH = -\log[H_3O^+] = -\log(1 \times 10^{-2}) = 2$$

Note the relationship between acidity and pH. As the acidity of the solution increases (concentration of hydronium ion increases) the pH decreases (see figure). A similar relationship exists for basic solutions. When a base is added to water the concentration of hydroxide ions increases and the hydronium ion concentration is lowered:

$$base + H_2O \rightarrow [H_3O^+]\downarrow + [\mathbf{OH^-}]\uparrow$$

Since pH always describes the hydronium ion concentration, if the concentration is now 1×10^{-12}M, then the pH would represent a basic solution (see figure):

$$pH = -\log[H_3O^+] = -\log(1 \times 10^{-12}) = 12$$

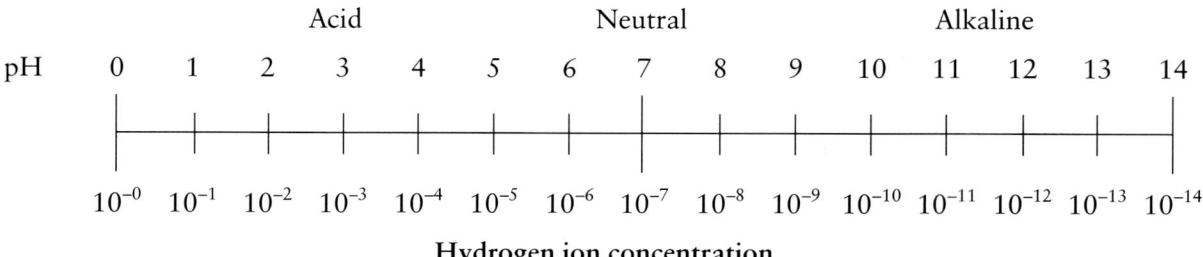

It is important to consider that the concentrations of hydronium and hydroxide ions in water are linked together; as the concentration of one increases the other must decrease. This means that they are always in equilibrium (balanced). Although the concentration values may increase or decrease, the product of the concentrations is always a constant value (the equilibrium constant for pure water: 1×10^{-14})

$$[H_3O^+] \times [OH^-] = 1 \times 10^{-14}$$

To summarize

- In neutral solutions $[H_3O^+] = [OH^-] = 1 \times 10^{-7}$ M and pH = 7
- In acidic solutions $[H_3O^+] > [OH^-]$ and $[H_3O^+] > 1 \times 10^{-7}$ M and pH < 7
- In basic solutions $[OH^-] > [H_3O^+]$ and $[H_3O^+] < 1 \times 10^{-7}$ M and pH > 7

Acid-base indicators are usually complex organic molecules that may behave as acids or bases. Molecules exhibiting this behavior are known as amphiprotic substances. An indicator will exhibit different colors depending on its amphiprotic nature. For example, phenolphthalein is an organic molecule. When it is exposed to a basic environment it behaves as an acid (donating a hydrogen ion) and turns pink.

$$\text{base } + \text{ indicator } \rightarrow \text{[indicator-H]}^-$$

(red)

The opposite reaction occurs when it is exposed to an acidic environment. The indicator acts as a base (accepting a hydrogen ion) and becomes colorless.

$$\text{acid } + \text{ indicator } \rightarrow \text{[H+indicator]}^+$$

(colorless)

Each indicator gains or loses a hydrogen ion within a very specific pH range. Some of the common indicators, pH ranges, and colors in acidic and basic solutions include:

Indicator	pH range	Color in acid	Color in base
Methyl orange	3.1–4.4	red	yellow
Congo red	3.0–5.2	blue	red
Chlorphenol red	5.2–6.8	yellow	red
Bromthymol blue	6.0–7.6	yellow	blue
Phenol red	6.8–8.4	yellow	red
Cresol purple	7.4–9.0	yellow	purple
Phenolphthalein	8.3–10.0	colorless	red

Appendix 3: List of abbreviations

abs	absolute	°F	degrees Fahrenheit	μg	microgram
aq	aqueous	ft	foot	N	normality, as in 1N
bp	boiling point	g	gram	p	pico (one trillionth or 10^{-12})
°C	degrees centigrade	hr	hour		
[C]	concentration	in.	inch	pH	acid base scale
ca.	approximately	insol	insoluble	ppm	parts per million
cc	cubic centimeter	k	karat	ppt	precipitate
cm	centimeter	L	liter	satd	saturated
concd	concentrated	log	logarithm	soln	solution
d	density	mass	mass	sp gr	specific gravity
deg or °	degree	M	molarity, mol/L	temp	temperature
dil	dilute	m	meter	V	volt
dild	diluted	mg	milligram	vol	volume
diln	dilution	mL	milliliter	v/v	volume per volume
e.g.	for example	mol	mole	wt	weight
eq	equation	mmol	millimole	w/v	weight per volume
et al.	and others	mp	melting point	w/w	weight per weight
etc.	and so forth	MW	molecular weight		

Appendix 4: Table of reagents with safety information used in spot tests

Based on MSDS information and the J. T. Baker SAF-T-DATA™ Ratings

Chemical names	Other names	Hazard class	Health rating	Contact rating	Signal	SAF-T-DATA™
Acetic acid, glacial		toxic, corrosive, flammable	moderate	severe (corrosive)	danger	white
Acetone		irritant, flammable	slight	slight	danger	red
Alizarin red S		irritant	moderate	severe	warning	orange
Aluminon	Aurintricarboxylic acid	irritant	slight	moderate	warning	orange
Ammonium acetate		irritant	slight	slight	caution	orange
Ammonium hydroxide	Ammonia solutions (10–35%)	toxic, corrosive	severe (poison)	severe (corrosive)	danger	white
Ammonium molybdate		irritant	moderate	moderate	warning	orange
Ammonium thiocyanate		irritant	moderate	slight	warning	orange
Anline		toxic, carcinogen flammable	severe (cancer)	severe (life)	danger	red
Ascorbic acid	Vitamin C	mild irritant	slight	slight	caution	orange
Barium chloride		toxic	severe (poison)	slight	danger	blue
Barium hydroxide		toxic	severe (life)	slight	danger	blue

slight < moderate < severe < extreme
caution: careful attention to avoid danger or harm; **warning**: advice to beware; **danger**: exposure or vulnerability to risk or peril
blue: health; **red**: flammable: **white**: corrosive: **yellow**: reactive; **orange**: general
*Prohibited in the UK.

Chemical names	Other names	Hazard class	Health rating	Contact rating	Signal	SAF-T-DATA™
Benzidine*		highly toxic, carcinogen	severe (cancer)	severe (life)	danger	blue
Cacotheline		toxic	slight	slight	caution	orange
Calcium hydroxide	Slaked lime	irritant, corrosive	slight	moderate	danger	orange
Calcium oxalate		irritant	slight	slight	caution	orange
Calcium oxide	Quicklime, unslaked lime	irritant, corrosive	slight	moderate	danger	orange
Cresol red		irritant	moderate	moderate	caution	orange
Cupric sulfate	Copper(II) sulfate, Blue vitriol	irritant	moderate	moderate	warning	orange
Dimethylamino-benzalderhyde	p-dimethylamino-benzaldehyde	irritant	slight	slight	caution	orange
Dimethyglyoxime	Diacetyldioxime	irritant	moderate	slight	warning	orange
Diphenylamine		irritant	slight	moderate	warning	orange
Diphenylecarbazide	1,5-diphenylcarbazide	irritant	slight	moderate	warning	orange
Diphenylcarbazone	s-diphenylcarbazone	irritant	slight	slight	caution	orange
Ethyl acetate		irritant, flammable	moderate	moderate	warning	red
Ethyl alcohol	Ethanol	irritant, flammable	slight	slight	caution	red
Ferric chloride	Iron(III) chloride	corrosive	moderate	severe (corrosive)	danger	white
Ferric sulfate	Iron(III) sulfate	irritant	slight	moderate	warning	orange
Ferrous sulfate	Iron(II) sulfate	irritant	slight	none	warning	orange
Glycerol	Glycerin	irritant	slight	slight	caution	orange
Hydrochloric acid	Muriatic acid	toxic, corrosive	severe (poison)	severe (corrosive)	danger	white
Hydrogen peroxide		mild irritant	none	slight	warning	orange
Hydroxylamine hydrochloride	Oxammonium hydrochloride	toxic, corrosive, possible mutagen	severe (poison)	severe (corrosive)	danger	white
Iodine		toxic, corrosive, oxidizer	moderate	severe (corrosive)	danger	white
Lead acetate	Lead(II) acetate	toxic, carcinogen, mutagen	severe (cancer)	slight	danger	blue
Mercuric chloride	Mercury(II) chloride	irritant, highly, toxic, mutagen	extreme (poison)	severe (life)	danger	blue
Mercuric thiocyanate	Mercury(II) thiocyanate	irritant, toxic, mutagen	severe (life)	severe (life)	danger	blue
Methyl alcohol	Methanol	toxic, flammable	severe (poison)	slight	danger	red
Methylene chloride	Dichloromethane	irritant, probable carcinogen, mutagen	severe (cancer)	moderate	warning	blue
Naphtol	1-naphtol, a-naphtol	irritant, toxic, oxidizer	severe	severe	danger	blue
Nitric acid		toxic, corrosive, oxidizer	severe (poison)	extreme (corrosive)	danger	yellow
o-toluidine	orthotoluidine	irritant, toxic, carcinogen, mutagen	severe (cancer)	severe	warning	blue

slight < moderate < severe < extreme
caution: careful attention to avoid danger or harm; **warning**: advice to beware; **danger**: exposure or vulnerability to risk or peril
blue: health; **red**: flammable: **white**: corrosive: **yellow**: reactive; **orange**: general
*Prohibited in the UK.

Chemical names	Other names	Hazard class	Health rating	Contact rating	Signal	SAF-T-DATA™
Phloroglucinol		irritant	slight	slight	warning	orange
Phosphoric acid		irritant, corrosive	moderate	severe (corrosive)	danger	white
Potassium dichromate		corrosive, oxidizer, carcinogen	extreme (cancer)	severe (life)	danger	yellow
Potassium ferrocyanide	Potassium hexacyanoferrate(II)	irritant	slight	slight	caution	orange
Potassium hydroxide	Caustic potash solution	toxic, corrosive	severe (poison)	severe (corrosive)	danger	white
Potassium iodide	Potide	irritant	moderate	moderate	caution	orange
Potassium permanganate		irritant, corrosive, oxidizer	moderate	severe (oxidizer)	danger	yellow
Silver nitrate		toxic, corrosive, oxidizer	severe (poison)	severe (corrosive)	danger	yellow
Sodium acetate		irritant	slight	slight	caution	orange
Sodium bisulfate	Sodium hydrogen sulfite	irritant	moderate	slight	warning	orange
Sodium chloride	Table salt	irritant	slight	slight	warning	orange
Sodium hydroxide	Caustic soda	corrosive, toxic	severe (poison)	extreme (corrosive)	danger	white
Stannous chloride	Tin(II) chloride dihydrate	irritant	moderate	moderate	warning	orange
Sulfuric acid	Oil of vitrol	corrosive, oxidizer, toxic	severe (poison)	extreme (corrosive)	danger	white
Sulfurous acid	Sulfur dioxide solution	corrosive	severe (poison)	severe (corrosive)	danger	white
Triglyceride (GPO Trinder)		irritant, toxic, mutagen	severe	extreme (life)	danger	blue
Triphenyltetrazolium chloride, tpt	TPTZ, TTC	irritant	moderate	moderate	warning	orange
Zinc metal granules or powder		irritant	slight	slight	warning	orange

slight < moderate < severe < extreme
caution: careful attention to avoid danger or harm; **warning**: advice to beware; **danger**: exposure or vulnerability to risk or peril
blue: health; **red**: flammable: **white**: corrosive: **yellow**: reactive; **orange**: general
*Prohibited in the UK.

Appendix 5: Chemicals, equipment and supplies

Chemicals

Acetic acid, glacial (liquid)
Acetone (liquid)
Alizarin Red S (solid)
Aluminon (solid)
Ammonium acetate (solid)
Ammonium hydroxide (28–30%, liquid)
Ammonium molybdate (solid)
Ammonium thiocyanate (solid)
Aniline (solid)
Ascorbic acid (Vitamin C; solid)
Barium chloride (solid)
Barium hydroxide (solid)
Benzidine (solid) (available only under special circumstances in the UK)
Cacotheline (solid)
Calcium hydroxide (solid)
Calcium oxalate (solid)
Calcium oxide (solid)
Cresol red (solid)
Cupric sulfate (Copper(II) sulfate, blue vitriol; solid)

Dimethylaminobenzaldehyde (p-Dimethylaminobenzaldehyde; solid)
Dimethylglyoxime (Diacetyldioxime; solid)
Diphenylamine (solid)
Diphenylcarbazide (1,5-Diphenylcarbazide; solid)
Diphenylcarbazone (s-Diphenylcarbazone; solid)
Diphenylthiocarbazone (Dithizone; solid)
Dithioxamide (Rubeanic acid; solid)
Ethyl acetate (liquid)
Ethyl alcohol (liquid)
Ferric chloride (Iron(III) chloride; solid)
Ferric sulfate (Iron(III) sulfate; solid)
Ferrous sulfate (Iron(II) sulfate; solid)
Glycerol (Glycerin; liquid)
Hydrochloric acid (liquid)
Hydrogen peroxide (Hydrogen peroxide solution; liquid)
Hydroxylamine hydrochloride (solid)
Iodine (solid)
Lead acetate (Lead(II)acetate; solid)

Mercuric chloride (Mercury(II) chloride; solid)
Mercuric thiocyanate (Mercury(II) thiocyanate; solid)
Methyl alcohol (Methanol; liquid)
Methylene chloride (Dichloromethane; liquid)
Napthol (1-Napthol, α-Napthol; solid)
Nitric acid (liquid)
o-Toluidine (liquid)
Phloroglucinol (solid)
Phosphoric acid (liquid)
Potassium dichromate (solid)
Potassium ferrocyanide (Potassium hexacyanoferrate(II); solid)
Potassium hydroxide (solid)
Potassium iodide (Potide; solid)
Potassium permanganate (solid)
Silver nitrate (solid)
Sodium acetate (solid)
Sodium bisulfite (solid)
Sodium chloride (Table salt; solid)
Sodium hydroxide (Caustic soda; solid)
Stannous chloride (Tin(II) chloride dihydrate; solid)
Sulfuric acid (liquid)
Sulfurous acid (liquid)
Triglyceride (GPO-Trinder) (liquid)
Triphenyltetrazolium chloride (solid)
Zinc metal granules or powder (solid)

Chemical test papers

- pH indicator paper
- pH pencil
- pH pen
- Reagent spot-test papers (antimony, arsenic, cuprotesmo, lead acetate, nickel, nitrate ions, phosphate ions, plumbtesmo, sulfate ions)

Equipment

- Burner (alcohol lamp, Bunsen burner, butane burner)
- Hot plate
- Heat source (lab oven, infrared lamp)
- Hood (chemical fume hood, fume absorber, fume scrubber, fume extractor, glove box, fan)
- Melting point apparatus (e.g. Mel-temp™ apparatus, or ask a local chemistry department)
- Micro centrifuge (e.g. Capsulefuge™)
- Balance; 1g, 0.1g, 0.01g (scale, balance: analytical, triple beam, hanging pan)

Glassware

- Beakers
- Capillary melting point tubes (1–2mm internal diameter)
- Capillaries (micro–mL capillaries or micro pipettes open on both ends) w/ micro pipette rubber bulb assembly
- Flasks (Erlemeyer 300mL)
- Glass stirring rods
- Graduated cylinders (various sizes)
- Microscope slides (glass with frosted end)
- Microscope cover slips (glass)
- Pipettes or droppers (plastic or glass, as fine-tipped as possible) (Pasteur)
- Reagent containers or bottles (clear glass) (dark glass) (acid resistant)
- Spot-test plates (glass-clear) (ceramic-white) (ceramic-black or dark blue)
- Test tubes (micro) (regular 10mL) (with stopper or screw cap – 50mL)
- Watch glass
- Vials (borosilicate glass; screw cap – flat bottom, 3mL)

Hand tools

- Brushes (for cleaning bottles, test tubes, pipettes)
- Lighter (butane, matches, flint)
- Magnification (magnifying glass, loupe, OptiVISOR®, stereo-zoom microscope)
- Microspatula (stainless steel)
- Mortar and pestle (glass, porcelain, agate)
- Scalpel blades (10 or 15 blade size)
- Scalpel handles
- Scissors (fine)
- Scoop (spatula, spoon)
- Test tube clamps or holders
- Thermometer (laboratory)
- Tongs (to hold hot beakers or crucibles)
- Tweezers or forceps (regular) (coated) (plastic) (stainless steel)
- Watch (timepiece with second hand, stopwatch, timer)

Supplies

- Centrifuge capsule tubes (size and type to fit the centrifuge)
- Batting (100% polyester spunbonded non-woven)
- Blotting paper (unbuffered)
- Plastic transfer pipettes
- Filter paper (qualitative and quantitative)
- Wrapping film (DuraSeal™) (Parafilm™)
- Swabs (applicators, Q-tips™, cotton tipped wooden sticks)
- Weighing accessories (paper, dishes, boats)
- Syringes

Set-ups

- **Electrolysis set-up:** (dry 6V battery) (electrolysis cables *or* four alligator clips and 100cm copper wire, 16 gauge insulated wire)
- **Radioactivity film set-up:** (Kodak T-Max® 400 film) (film bag/light proof plastic) (processing chemicals)
- **Pencil hardness set-up:** (Pencil set 6B – HB – F– H6) (eraser for pencil marks – vinyl, rubber, gum)
- **Jeweler's touchstone with karat testing needles**
- **Specific gravity set-up:** (Foam board, expanded polyethylene foam, aluminium foil, wire)

Appendix 6: Product suppliers

Safety product suppliers

Fisher Scientific Safety Products
2761 Walnut Ave.
Tustin, CA 92681
1-800 926-8999
http://www.fisher1.com/

J. T. Baker Safety Products
222 Red School Lane
Phillipsburg, NJ 08865
1-800 582-2537
http://www.jtbaker.com/ps/ps.htm

Lab Safety Supply Co.
P.O. Box 1368
Janesville, WI 53547
1-800 356-0783
http://www.labsafety.com/

Chemical suppliers

Aldrich Chemical Company, Inc.
1001 W. Saint Paul Ave.
Milwaukee, WI 53233
1-800 588-9160
http://www.sigald.sial.com/aldrich/aldrich.html

Chem Service, Inc.
P.O. Box 599
West Chester, PA
19381-0599
1-800 452-9994 (orders only)
(610) 692-3026 (telephone)
(610) 692-8729 (FAX)
http://chemservice.com
info@chemservice.com

Fisher Scientific (Chemical Catalog)
711 Forbes Avenue
Pittsburgh, PA
15219-4785
1-800 766-7000
http://www.fisher1.com/

Fluka Chemie AG
Industriestrasse 25
CH-9471 Buchs, Switzerland
1-800 358-5287
http://www.sigald.sial.com/
fluka/fluka.html

Hach Company
P.O. Box 389
Loveland, CO 80539
1-800 227-4224
http://www.hach.com/

Sigma Chemical Company
P.O. Box 14508
St. Louis, MO 63178
1-800 325-3010
http://www.sigald.sial.com/
sigma/sigma1a.htm

Thomas Scientifc
P.O. Box 99
Swedesboro, NJ 08085
1-800 345-2100 x. 6612
http://www.techexpo.com/
firms/thomscie.html

VWR Scientific
P.O. Box 6016
Cerritos, CA 90702
1-800-289-8972
http://www.vwrsp.com/

Test paper supplier

Gallard-Schlesinger Industries, Inc.
584 Mineola Avenue
Carle Place, NY 11514
1-516 333-5600

EM Science
480 S. Democrat Road
Gibbstown, NJ 08027
1-800 222-0342
http://www.emscience.com/

pH testing tools

Talas
568 Broadway
New York, NY 10012
1-212-219-0770

University Products, Inc.
517 Main Street
P.O. Box 101
Holyoke, MA 01041
1-800-532-9281
http://
www.universityproducts.com/

Metal testing supplies

Rio Grande
4516 Anaheim Ave. NE
Albuquerque, NM 87113
1-800-545-6566

Gesswein
255 Hancock Ave.
Bridgeport, CT 06605
1-800-243-4466

Norton Goldsmiths
Spot Testing Supplies
P.O. Box 43851
Tucson, AZ 85733

Small equipment

Laboratory Devices
Mel-Temp™
Box 6402
Holliston, MA 01746-6402
1-800-447-6722

Whatman Lab Sales
Capsulefuge™
PO Box 1359
Hillsboro OR 97123
1-800-942-8626

Local resources

- Jewelers' supply
- Lapidary supply
- Foundry companies
- Plating companies
- Salvage merchandise
- Electronics supply

Appendix 7: Materials Characterization Trial Form

Spot Test for Material Identification Form

		File name
Trial #	Test title	
Date	Sample source	
	Tester	

Reagent preparation
(Reagent weight, volumes, etc.)

Test data
(Methods, running time, samples, conditions, etc.)

Observations
(Positive, negative, colour change, sample change, odor, etc.)

Remarks
(Interpretation of results, problems, etc.)

Spot Test for Material Identification Form

File name

Test title Test for tin using cacotheline

Trial # 1

Sample source Plate 6208, pitcher 6225, knife E-7313C

Date 29 April 1996

Tester Scott Carroll

Reagent preparation *(Reagent weight, volumes, etc.)*	The saturated cacotheline solution was made by adding 15mL of distilled water to 1g of cacotheline powder. This made about a 6% solution. The saturated salt solution was made by adding salt to about 30mL of water until the salt no longer dissolved. This solution was left overnight to make sure that there were undissolved crystals remaining in the bottom.
Test data *(Methods, running time, samples, conditions, etc.)*	The objects are all historical objects from storage. The plate and pitcher appear to be made from pewter and the knife has a wooden handle and a blade that appears to be iron or steel. For this trial the electrolysis set up was used and not the alternative methods with acid. The electrolytic cell was set up as per the instructions for each object in succession. The filter paper was dipped in the cacotheline solution and allowed to dry. The paper was then wetted with the saturated salt solution and held by tweezers attached with an alligator clip to the negative pole of the battery. The positive pole was connected to the object with an alligator clip. The tip of the paper was touched to the object for about 20s in each case.
Observations *(Positive, negative, colour change, sample change, odor, etc.)*	The plate and the pitcher tested positive for tin. The test paper started showing purple for the plate very quickly. It took a couple of attempts before it showed purple for the pitcher. The knife did not appear to give a positive. On one attempt it did turn a darker color, but was not definitely purple.
Remarks *(Interpretation of results, problems, etc.)*	The fact that the plate and pitcher turned the test paper purple indicates that they do contain tin and probably are pewter, as originally thought. This will have to be confirmed by testing for lead in both of them. The fact that the knife did not turn the test paper purple indicates that there is no tin present and also shows that the test is working properly and did not give a false positive. It is tricky to get the right ratio of salt and cacotheline solution; in which case the test did not seem to work. The best results were obtained when the cacotheline solution was allowed to dry first, or slightly dry, before the salt (NaCl) was added.

Appendix 8: Glossary

Abrasive: A substance used to wear away another substance by friction.
Absorption: The penetration of molecules of one substance into the body of another substance.
Accuracy: The agreement between the result of a measurement and the true value of the quantity measured (not the same as precision).
Acid: (1) A substance that, when dissolved in pure water, increases the concentration of hydronium ions, H_3O^+ (Arrhenius definition); (2) any substance that donates a hydrogen ion to any other substance (Brønsted–Lowry definition); (3) A substance that can accept a pair of electrons to form a new bond (Lewis definition). See also *Organic acid* and *Mineral acid*.
Acute toxicity: The adverse (acute) effects resulting from a single dose of, or exposure to, a substance.
Adsorption: The process in which molecules of a liquid or a gas condense on the surface of a solid.
Aldehyde: A broad class of organic compounds containing the structural element –CHO.
Alkali: See *Base*.
Alloy: A compound metal resulting from melting two or more metals together, causing a change in properties.
Alum: A sulfate salt with the molecular formula, $KAl(SO_4)_2.12H_2O$; commonly associated with a monovalent and a trivalent metal; e.g. aluminum ammonium sulfate or aluminum potassium sulfate.
Amber: A transparent to translucent fossil resin derived from a variety of pine (sometimes containing insects or plants).
Amorphous: The characteristic applied to solid materials that have no apparent crystalline structure.
Anode: (1) The electrode of either an electrolysis cell or an electrochemical (voltaic

or battery) cell at which oxidation occurs; (2) in an electrolysis cell, the positive (+) electrode or terminal; (3) in a voltaic cell, the negative (–) electrode or terminal.

Aqua regia: A solution of nitric acid (HNO_3) and hydrochloric acid (H_2SO_4) in a ratio of approximately 1:3 or 1:4 (v:v).

Aqueous (*aq*): A material dissolved in water.

Assay: The analysis of an ore or alloy to determine constituents and their proportions.

Balance: An instrument used for determining metric weight of materials accurately.

Base: (1) A substance that, when dissolved in pure water, increases the concentration of hydroxide ions, OH^- (Arrhenius definition); (2) any substance that accepts a hydrogen ion from any other substance (Brønsted–Lowry definition); (3) a substance that can donate a pair of electrons to form a new bond (Lewis definition).

Boiling point: The temperature at which the vapor pressure of a liquid is equal to the external (atmospheric) pressure.

Buffer: A solution that resists a change in pH upon addition of hydroxide (OH^-) or hydronium (H_3O^+) ions.

Capillary tube: A small slender glass form or tube having a very small bore. Open-ended tubes may be used with a bulb like a pipette; closed-ended tubes may be used with a melting point apparatus.

Carat: A unit of weight for diamond and other gemstones. The metric carat (200 milligrams) is now the standard in most countries. In some countries carat is synonymous with karat (see *Karat*).

Carbohydrate: A compound that contains carbon, hydrogen and oxygen in a 1:2:1 ratio. Carbohydrates comprise (1) monosaccharide or simple sugars such as fructose and glucose; (2) disaccharides such as sucrose and lactose as well as polysaccharides such as starch, cellulose, and natural gums.

Carbonize: To convert into carbon or carbonic residue, usually by heating.

Carcinogen: A substance determined to be cancer producing or potentially cancer producing.

Cathode: (1) The electrode of either an electrolysis cell or an electrochemical (voltaic or battery) cell at which reduction occurs; (2) in an electrolysis cell, the negative (–) electrode or terminal; (3) in a voltaic cell, the positive (+) electrode or terminal.

Centrifuge: A device for separation of solid materials from suspension by whirling at high speeds.

Chemical equation: A written representation of a chemical reaction, showing the reactant and products, their physical states (e.g. (*g*), (*l*), (*s*), (*aq*)), and the direction in which the reaction proceeds. A 'balanced' chemical equation also includes the exact molar ratio of reactants and products.

Coagulation: The clumping of fine particles together that give larger particles to facilitate separation or settling or filtration.

Combustible: A term used to classify liquids that will burn, on the basis of their flash points in a specified test. Non-liquids like wood and paper are classified as ordinary combustibles.

Concentrated solution: A solution containing a large concentration of solute dissolved in a solvent.

Copper: Cu, a malleable, durable and distinctively reddish metal. Interesting alloys include: brass (Cu : Zn); bronze (Cu : Sn); nickel silver or German silver (Cu : Ni : Zn).

Corrosive: A chemical that causes visible destruction or irreversible alterations by chemical action at the site of contact.

Dilute solution: A solution with a small amount of solute dissolved in a solvent.

Distillation: The process in which a liquid is vaporized and then condensed in order to remove impurities.

Effervescence: The escape of gas from a liquid due to a decrease in pressure or an increase in temperature that is often characterized by bubbling or hissing.

Electrolysis: The use of electrical energy to produce a chemical reaction.

Endothermic: A chemical reaction that requires heat in order to proceed.

Exothermic: A chemical reaction that is accompanied by the evolution of heat.

Explosive: A material that causes a sudden, almost instant release of pressure, gas, and heat when subjected to sudden shock, pressure, or high temperature.

Filter paper: The material used to separate suspended solids from a liquid by allowing the mixture to flow through the porous paper barrier. (1) Qualitative papers will leave an appreciable amount of ash upon ignition but are generally used for the filtration of precipitates. (2) Quantitative papers are more pure and do not produce an ash that could interfere when precipitates are to be ignited on the paper and weighed. (3) Hardened papers are designed for use in vacuum filtration and are made to have greater wet strength and lintless surfaces.

Flammable: Any solid, liquid, vapor, gas that will ignite easily and burn rapidly. (1) A solid, other than an explosive that will cause fire through friction, absorption of moisture, spontaneous chemical change, or when ignited will burn so vigorously and persistently as to create a serious hazard. (2) A liquid having a flashpoint below 37.8°C (100°F). (3) A gas that at ambient temperature and pressure, forms a flammable mixture with air at a concentration of 13% by volume or less, or a gas that forms a range of flammable mixtures with air wider than 12% by volume at ambient temperature and pressure. (4) A vapor or aerosol that yields a flame when ignited.

Flashpoint: The lowest temperature at which a liquid or volatile solid gives off a vapor, sufficient to form an ignitable mixture with the air near the surface of the liquid within the test vessel.

Gas (g): A state of matter that is unrestricted by cohesive force and having no definite volume or shape.

Glacial: The term applied to a number of acids (e.g. acetic (CH_3COOH) and phosphoric (H_3PO_4)) which are appreciably free of water and have a freezing point slightly below room temperature when in a highly pure state. For example glacial acetic acid is 99.8% pure and crystallizes at 16.6°C (62°F).

Gold: Au, a dense lustrous yellow metal. Some interesting gold alloys include: British standard gold (92Au : 8Cu); 14k (58Au : 14–28Cu: 4–8Ag); pink gold (50Au : 50Cu); white gold (75–85Au : 8–10Ni : 2–9Zn).

Grade: A reference to purity standards for chemicals and chemical products established by various specifications.

Hazardous material: Any material or substance that can be damaging to the human health if not handled properly. A broad range of materials may be classified as (1) irritant or harmful allergen, (2) toxic or poisonous, (3) corrosive on contact or inhalation, (4) flammable, (5) explosive, (6) radioactive, or (7) biohazard.

Hydrolysis reaction (hydrolyze): A reaction with water in which a bond to oxygen is broken, often resulting in the formation of two products from one reactant. For example, the hydrolysis of sucrose yields a molecule of glucose and a molecule of fructose by reaction with water in the presence of an enzyme or acid catalyst.

Interference: An obstruction that causes the results of a chemical reaction to be augmented, diminished, or otherwise affected.

Ion: An atom, radical, or molecule that has lost or gained one or more electrons, and has thus acquired an electric charge. Positively charged ions are cations, and negatively charged ions are anions. Ions have different properties from the elements that form them. When ions are dissolved in water they are known as electrolytes (an electrolyte solution), owing to their ability to conduct electricity.

Irritants: Chemicals which are not corrosive, but which cause a reversible inflammatory affect on living tissue by chemical action at the site of contact.

Karat: A unit of weight to express the proportion of gold in an alloy or the quality of a gold alloy. Pure gold is 24 karats. Gold alloys are known by their proportion in weight of fine gold; e.g. 18 karat is 18/24 fine gold or 75% of the weight of the entire article. See also *Carat*.

Loupe: Any small magnifying glass mounted for use in the hand as a *hand loupe*, held in the eye socket or attached to spectacles as an *eye loupe*, or worn on the head as a *head loupe*.

Mass: The measure of quantity of matter. The term is used in discussing measurements made with a balance. The common metric (also known as the International System or SI) units of mass are grams (g) and kilogram (kg).

Mel-temp® apparatus: An instrument that measures melting point or the temperature at which the crystals of a pure substance are in equilibrium with the liquid phase at atmospheric pressure.

Meniscus: The curved surface of a liquid within a container caused by surface tension.

Microscope: An optical instrument that affords magnification of objects or particles.

Mineral acid: A large class of substances derived from a wide range of non-carbon containing (or inorganic) elements: hydrochloric (HCl), sulfuric (H_2SO_4), nitric (HNO_3).

Mixture: A combination of two or more substances in which each substance retains its physical state, and which theoretically can be separated into the pure substances.

Moh's scale: The most commonly used non-linear scale that characterizes the relative hardness of minerals. The scale includes unequal units identified in divisions as 1 to 10.

Molar solution: A molar solution contains 1 mole of a substance per liter of solvent.

Molarity: The concentration of a solution, expressed as moles of dissolved substance per liter of solution.

Mole: SI unit of the amount of a substance, defined as the amount of a substance that contains as many elementary entities (atoms, molecules, etc.) as there are atoms in 12g of the isotope carbon-12.

MSDS: Materials Safety Data Sheets are a Federally mandated form supplied by vendors of chemicals in the United States describing their chemical's potential hazard and information on how people should protect themselves from these hazards. (See section 2.1.)

Mutagen: A chemical or physical agent that induces genetic mutations.

Nickel: Ni, a malleable, silvery metal. Interesting alloys include: US nickel coins (75Cu : 25Ni); nickel silver (46Cu : 34Zn : 20Ni); monel (68Ni : 29Cu : 1Fe : 1Mn : other).

Opti-visor®: A head loupe or simple binocular magnifier that is attached to a head frame.

Organic acid: Carboxyl-containing substances having the functional group –COOH.

Oxidizing agent: Any substance that accepts electrons either at room temperature or under slight heating and is reduced in an oxidation-reduction reaction. The term includes such chemicals as peroxides, chlorates, perchlorates, nitrates, and permanganates. They can react violently at room temperatures when stored near or in contact with reducing materials such as cellulosic and other organic compounds. Storage should be well ventilated and kept as cool as possible.

Parafilm: Trademark for a wrapping film, Parafilm M®, made by the American Can Co.

PEL: Permissible Exposure Limit, a term used by the Occupational Safety and Health Administration (OSHA) of the United States to express the legal airborne concentration of a material to which persons can be exposed day after day. Usually stated in parts per million (ppm) or milligrams per cubic meter (mg/M^3).

Percent solution: The number of grams of solute per 100 grams of solution (or the mass of solute divided by the mass of solution and multiplied by 100).

Pewter: A white metal; a tin (Sn) alloy with 5–15% antimony (Sb), 0–3% copper (Cu), 0–15% lead (Pb). Similar metals include Britannia metal (Sn : Sb) and white metal (Sn : Sb : Cu).

pH: A measure of acidity or basicity of a solution. The negative of the base –10 logarithm of the hydronium ion concentration ($-\log[H_3O^+]$).

pH paper: A type of paper that has been infused with pH indicators (usually organic dyes) that change color to indicate the approximate pH of a solution.

Polymerization: A chemical reaction, usually carried out with a catalyst, heat or

light, or pressure, in which a large number of relatively simple molecules combine to form a chain-like or network macro-molecule.
Polysaccharide: A polymeric chain of monosaccharides or simple sugars (e.g. glucose, fructose). Cellulose and starches are examples of polysaccharides.
Precision: A statistical value based on the degree of agreement of repeated measurements of the same quantity (not the same as accuracy).
Pyrolysis: The transformation of a compound into one or more other substances by heat alone (burning).
Pyrophoric: A chemical that will ignite spontaneously or explode in air at a temperature of ≤54.4°C (130°F).
Qualitative analysis: The identification of materials based on non-numerical experimental observation of chemical and physical properties.
Quantitative analysis: The identification of materials based on numerical experimental observation of chemical and physical properties such as the percentage composition of mixtures.
Reagent: Any substance capable of undergoing a reaction with another substance or itself.
Reducing agent: Any substance that donates electrons and is oxidized in an oxidation-reduction reaction.
Reducing sugar: Sugars such as glucose that will reduce Fehling's solution or similar test reagents.
Salt: Any substance that is comprised of a cation(s) and anion(s) as part of its molecular formula; also referred to as an *ionic compound*.
Saturated solution: A solution that contains the maximum amount of solute at a given temperature.
Sensitivity: The change in quantity required to produce a perceptible change in indication. It is not a term to indicate the quality of a measurement.
Sensitizers: Chemicals that cause a substantial proportion of exposed people or animals to develop an allergic reaction in normal tissue after repeated exposure to the chemical.
Silver grades: A soft, lustrous white metal that is available as: (1) pure or fine, (2) sterling –92.5%, (3) coin –90% (in the United States), and (4) German –0%.
Solid (*s*): The state of matter having definite volume, shape and mass.
Solution: The homogeneous mixture of two or more substances.
Specific gravity: The ratio of the mass of a given volume of a substance to the mass of an equal volume of water at 4°C. The temperature of the measurement must be stated. Also known as specific weight.
Standard solution: A solution with a precisely known composition and concentration of solute that is used to determine the concentration of another solution.
Supernatant: The clear liquid or fluid that forms as a layer on the surface of another liquid or solid.
Supersaturated solution: A solution which contains more dissolved solute than it could normally hold at a given temperature and pressure.
Teratogen: An agent or substance that causes physical defects in a developing embryo.
TLV: Threshold Limit Values, a term used by the American Conference of Government Hygienists (ACGIH) to express the airborne concentration of a material to which nearly all persons can be exposed day after day, without adverse health effects.
Touchstone: A smooth, hard, black stone (traditionally quartz or jasper), used to detect differences between an unknown metal and a test needle (set of metals of known quality) by comparing the chemical reactions of streaks made on the stone surface.
Toxic: Chemicals that cause damage to living tissue by ingestion, inhalation, or absorption through the skin. Toxicity levels are based on test dosages made on experimental animals under controlled conditions. Most important of these are the LD_{50} (lethal dose 50%) and LC_{50} (lethal concentration 50%) tests.

Tweezers: A tool for quick and efficient handling of small parts consisting of a pair of spring-metal arms that are fastened together at one end and worked by the fingers. Sometimes called forceps.

Ultraviolet lamp: A source of ultraviolet radiation used for exciting fluorescence in materials. The lamps include long wave and short wave types.

Unstable: A reactive chemical that will vigorously polymerize, decompose, or condense under conditions of shocks, pressure, temperature.

Water purification: (1) Boiled: a purification process that involves agitation, separation, and removal of carbon dioxide. (2) De-ionized: a purification process that involves removing ions. (3) Distilled: a purification process that involves separation of components by converting the liquid into a vapor and then condensing the vapor to a liquid.

Weight: Weight equals mass times the gravitational attraction. Mass is proportional to weight, but the unit of weight is the newton.

Appendix 9: Bibliography

+ denotes references cited in this volume

+ American Society for Testing and Materials. 1989. Standard Test Method for Film Hardness by Pencil, Test # D 3363-74, reapproved. *Annual Book of Standards,* vol. 6.01, 518–19. West Conshohocken, PA: American Society for Testing and Materials.
+ Anft, Berthold. 1955. Friedlieb Ferdinand Ruge: A Forgotten Chemist of the Nineteenth Century. *Journal of Chemical Education* **32**(11): 566–74.
  Banik, Gerhard. 1983. *Scientific Principles of Conservation Course-Notes, Practical Exercise: Spot Tests.* Rome: ICCROM.
+ Bauer, Wilhelm P. 1983. *Class notes. Scientific Principles of Conservation.* Rome: ICCROM.
+ Beaubien, Harriet F. 1995. Low-tech Methods for Characterizing Materials in the Field. In *Materials Issues in Art and Archaeology* IV, 641–51. Eds P. B. Vandiver, J. R. Druzik, J. L. Galvan Madrid, I. C. Freestone and G. S. Wheeler. Materials Research Society Symposium Proceedings. Volume 352. Pittsburgh: Materials Research Society.
  Blackshaw, Susan M., and Vincent D. Daniels. 1979. Testing of Materials for Use in Storage and Displays in Museums. *The Conservator* **3**: 16–19.
+ Blount, Alice. M. 1990. A Low-cost Radioactivity Test for Geological Specimens. *Collection Forum* **6**(1): 8–11.
+ Borelli, E. 1993. *Mural Paintings: Conservation Course-Identification of Binding Media-Lab Notes,* Part I, *Constituent Materials/Execution Techniques.* Rome: ICCROM.
  Bradford, Marion M. 1976. A Rapid and Sensitive Method for the Quantitation of Microgram Quantities of Protein Utilizing the Principle of Protein-Dye Binding. *Analytical Biochemistry* **72**: 248–54.
+ Braun, Dietrich. 1982. *Simple Methods for Identification of Plastics.* New York: Macmillan.
+ Browning, B. L. 1977. *Analysis of Paper.* New York: Marcel Dekker.
+ Carroll, Scott, and Leo Kohn. 1994. Silicone vs. Rubber Gaskets. *Western Association for Art Conservation Newsletter* **16**(2): 19.

Chamot, Emile M. and Clyde W. Mason. 1982. *Handbook of Chemical Microscopy*. 4th edn. New York : Wiley-Interscience.

Coxon, Helen C. 1993. Practical Pitfalls in the Identification of Plastics. In *Saving the 20th Century; The Conservation of Modern Materials: Proceedings of a Conference, Symposium '91*. 395–409. D. Grattan, ed. Ottawa: Canadian Conservation Institute.

Eidt, Robert C. 1973. A Rapid Chemical Field Test for Archaeological Site Surveying. *American Antiquity* **38**(2): 206–10.

+ Fansett, George R. 1934. *Field Tests for the Common Metals*. (6th edn.). Arizona Bureau of Mines, Mineral Technological Series No. 36, Bulletin No. 136. Tucson: University of Arizona Press.

+ Feigl, Fritz. 1937. *Qualitative Analysis by Spot-tests; Inorganic and Organic Applications*. First English Edition with translation by J. W. Matthews. New York: Elsevier Publishing Co.

+ Feigl, Fritz, and Vinzenz Anger 1966. *Spot Tests in Organic Analysis*, 7th edn. New York: Elsevier.

+ Feigl, Fritz, and Vinzenz Anger. 1972. *Spot Tests in Inorganic Analysis*, 6th English edn. New York: Elsevier.

Feller, Robert L., ed. 1986. *Artists' Pigments: A Handbook of their History and Characteristics*, Vol 1. Washington, D.C.: National Gallery of Art.

+ Fenn, Julia. 1996. The Cellulose Nitrate Time Bomb: Using Sulphonephthalein Indicators to Evaluate Storage Strategies. In *From Marble to Chocolate: The Conservation of Modern Sculpture*, Jackie Heuman, ed. 87–92. London: Archetype.

Fitzhugh, Elisabeth West. 1997. *Artists' Pigments; A Handbook of Their History and Characteristics*, Vol. 3. Washington, D.C.: National Gallery of Art.

+ Florian, Mary-Lou. 1984. Analytical Methods for Protein Identification and Characterization; Simple Analytical Methods Using Case Studies to Illustrate Techniques, Section 5. In *Protein Chemistry for Conservators*, C. L. Rose and D. W. Von Endt, eds, 89–96. Washington, DC: American Institute for Conservation.

Found, Christine and Kate Helwig. 1995. The Reliability of Spot Tests for the Detection of Arsenic and Mercury in Natural History Collections: A Case Study. *Collection Forum* **11**(1): 6–15.

Ganiaris, Helen. 1985. A Portable Spot-Test Kit. *Conservation News* **27**: 25–6.

Gardiner, Joy W. 1988. Fabric Finishes: Tests and Consequences, Part 2. In *20th Century Materials, Testing and Textile Conservation: 9th Symposium of the Harpers Ferry Regional Textile Group*, 39–44. Washington, DC: American Institute for Conservation and the Harpers Ferry Regional Textile Group.

Gatenby, Sue. 1993. An Identification Method for Fat and/or Oil Binding Media Used on Australian Aboriginal Objects. In *ICOM Committee for Conservation Tenth Triennial Meeting, Washington, DC*, Janet Bridgland, ed. 167–71. Paris: International Committee of Museum, preprint.

+ Gedye, Ione, Henry Hodges and Andrew Oddy. 1973. *Notes for a Short Course in Conservation*. London: British Museum Research Laboratory.

Gettens, Rutherford J., and George L. Stout. 1966. *Painting Materials: A Short Encyclopaedia*. New York: Dover (original publication, 1942. NY: Van Nostrand Co.).

Goodyear, Frank H. 1971. *Archaeological Site Science*. New York: American Elsevier.

Gordon, Arnold J., and Richard A. Ford. 1972. *The Chemist's Companion: A Handbook of Practical Data, Techniques, and References*. New York: John Wiley & Sons.

Green, Lorna R. and David Thickett. 1995. Testing Materials for Use in the Storage and Display of Antiquities; a Revised Methodology. *Studies in Conservation* **40**(3): 145–52.

Hawley, Florence M. 1929. Prehistoric Pottery Pigments in the Southwest. *American Anthropologist* **31**(4): 731–45.

Hawley, Florence M. 1938. Classification of Black Pottery Pigments and Paint Areas. *University of New Mexico Bulletin* **321**: 3–11.

Hodges, Henry. n.d. Micro and Semi-micro Methods of Chemical Analysis. Notes from lectures made at the Institute of Archaeology, London and the Queens University, Kingston, Canada.

+ Hofenk-de Graaff, Judith H. 1974. A Simple Method for the Identification of Indigo. *Studies in Conservation* **19**(1): 54–5.

+ Hogness, Thorfin R., and Warren C. Johnson. 1954. *Qualitative Analysis and Chemical Equilibrium*, 4th edition. New York: Holt.

+ Hopwood, Walter R. 1979. *Choosing Materials for Prolonged Proximity to Museum Objects.* American Institute for Conservation, 7th Annual Meeting, Toronto, 44–9. Washington, DC: American Institute for Conservation, preprint.

+ Howie, Francis M. P. 1984. Materials Used for Conserving Fossil Specimens Since 1930: A Review. In *Adhesives and Consolidants*, N. S. Brommelle, Elizabeth M. Pye, Perry Smith and Garry Thompson, eds. 92–7. London: International Institute for Conservation, Paris Congress, preprint.

+ International Nickel Company, Inc. 1951. *Rapid Identification (spot testing) of Some Metals and Alloys.* New York, NY: Development and Research Division, International Nickel Company.

Jaffe, Harold. 1989–1990. Testing Metal Leaf Surfaces. The Gilder's Tip. *Newsletter of the Society of Gilders* 3(1): 4.

Jungreis, Ervin. 1985. *Spot Test Analysis; Clinical, Environmental, Forensic, and Geochemical Applications.* New York: John Wiley & Sons.

+ Knapp, Anthony M. 1993. Arsenic Health and Safety Update. *Conserv-O-Gram* No. 2/3. Harpers Ferry WV: Curatorial Services Division, National Park Service, US Department of the Interior.

+ Koob, Stephen P. 1982. The Instability of Cellulose Nitrate Adhesives. *The Conservator* 6: 31–4.

Krause, A., and A. Lange. 1969. *Introduction to the Chemical Analysis of Plastics.* London: Iliffe Books.

Kunkel, Henry G., and Arne Tiselius. 1951. Electrophoresis of Proteins on Filter Paper. *Journal of General Physiology* 35(1): 89–118.

Larson, Renee. 1990. Micro-Chemical Determination of Vegetable Tannins. *Leather Conservation News* 7(1): 6–7.

+ Laver, Marilyn. 1978. Spot Tests in Conservation: Metals and Alloys. *International Committee of Museums (ICOM), Committee for Conservation*, 5th Triennial Meeting, Zagreb, 78/23/8: 1–11, preprint.

Lawrence, Sandra. 1976. *A Review and Experimental Evaluation of Microchemical Tests for Some White Pigments.* Masters thesis (unpublished), Queen's University, Kingston, Ontario, Canada.

Lee, L. R., and D. Thickett. 1996. *Selection of Materials for the Storage or Display of Museum Objects.* British Museum Occasional Paper No. 111. London: British Museum Press.

+ Liddicoat, Richard. T., Jr., and Lawrence L. Copeland. 1976. *The Jewelers' Manual.* Los Angeles: Gemological Institute of America.

+ Macherey-Nagel & Co. 1982. Nickel Test Paper for the Rapid Determination of Nickel in Solutions and in Nickel-containing Alloys. Product Instructions No. Ni/01/0/2.82. Dueren, Germany: Machery-Nagel & Co.

Mantell, C. L., C. W. Kopf, J. L. Curtis and E. M. Rogers. 1942. *The Technology of Natural Resins.* New York: John Wiley & Co.

Marshall, Joan A. 1992. *The Identification of Flax, Hemp, Jute and Ramie in Textile Artifacts.* Masters thesis (unpublished), University of Alberta, Edmonton, Alberta, Canada

+ Mayer, Debora D., compiler. 1990. Spot Tests. In *Paper Conservation Catalog*, 7th edn. 21–2. Washington DC: American Institute for Conservation Book and Paper Group.

Mellor, Joseph W. 1923. *A Comprehensive Treatise on Inorganic and Theoretical Chemistry*, Vol. 4. London: Longmans, Green.

+ Merck KGaA, 1990. Merckoquant Sulfate Test Paper. Product Instructions. No. 1.10019.0001. Darmstadt, Germany: Merck.

+ Merck KGaA, 1990. Merckoquant Nitrate Test Paper. Product Instructions. No. 1.10020.0001. Darmstadt, Germany: Merck.

+ Moss A. A. 1956. *The Identification of Metals. Handbook for Museum Curators*, Part B, *Museum Technique*, Section 8: 2–8. London: Museums Association.

National Research Council. 1995. *Prudent Practices in the Laboratory.* Washington DC: National Academy Press.

Nelson, Fred W. 1975. A Simple Method for Distinguishing Between Organic and Inorganic Paints on Black-on-white Anasazi Pottery. *American Antiquity* 40(3): 348–9.

Nicholson, Catherine, and Elissa O'Loughlin. 1996. The Use of A-D Strips for Screening Conservators and Exhibit Materials. In *Book and Paper Group Annual*, Volume 15. Robert Espinosa, compiler. 83–5. Washington DC: American Institute for Conservation.

Organ R. M. 1967. Spot-Tests for Application Directly to Metals. *American Institute for Conservation Bulletin* **10**(1): 17–19.

Pavia, Donald L., Gary M. Lampman, and George S. Kris. 1982. *Introduction to Organic Laboratory Techniques: A Contemporary Approach,* 2nd ed. Saunders Golden Sunburst Series. Philadelphia: Saunders College Publications.

Pearson, Colin. 1983. Unpublished Lecture Notes: Classnotes from the Inorganic Materials Course. Available from the author at the National Center for the Conservation of Cultural Property. University of Canberra, Australia.

+ Persson, K. B. 1997. Soil Phosphate Analysis: A New Technique for Measurement in the Field Using a Test Strip. *Archaeometry* **39**(2): 441–3.

+ Reed, R. 1972. *Ancient Skins, Parchments and Leather.* London: Seminar Press.

Reilly, Julie. 1991. Celluloid Objects: Their Chemistry and Preservation. *Journal of the American Institute of Conservation* **30**(2): 145–62.

+ Rémillard, France. 1995. *La Preservation des Objects de Plastique et de Caoutchouc. Seminaire de Formation.* Quebec: Centre de Conservation du Quebec.

Roy, Ashok, editor. 1993. *Artist's Pigments; A Handbook of their History and Characteristics,* Vol. 2. Washington, DC: National Gallery of Art.

Sax, N. Irving and R.J. Lewis. 1996. *Sax's Dangerous Properties of Industrial Materials*, 7th edn. New York: Van Nostrand Reinhold.

+ Schramm, Hans-Peter. 1995. *Historische Malmaterialien und Ihre Identifizierung.* Stuttgart: Ferdinand Enke Verlag.

Selwyn, Lyndsie. 1993. Course Notes from What's New in Old Metals? Specialized seminar. Available from author at the Canadian Conservation Institute, Ottawa.

Semczak, Carl M. 1977. A Comparison of Chloride Tests. *Studies in Conservation* **22**(1): 40–1.

+ Sorum C. H. 1960. *Introduction to Semimicro Quantitative Analysis,* 3rd edn. Englewood Cliffs, N.J.: Prentice-Hall.

+ Stulik, Dusan, and Henry Florsheim. 1992. Binding Media Identification in Painted Ethnographic Objects. *Journal of the American Institute for Conservation* **31**(3): 275–88.

+ Tabasso M. Laurenzi. 1993. *Mural Paintings Conservation Course — Identification of Pigments — Lab Notes,* Part I: *Constituent Materials/Execution Techniques.* Rome, Italy: ICCROM.

+ Tetreault, Jean. 1992. La Mésure de L'acidite des produits volatils (Measuring the Acidity of Volatile Products). *Journal of International Institute for Conservation-Canadian Group* **17**: 17–25.

+ Untracht, Oppi. 1982. *Jewelry Concepts and Technology.* Garden City, New York: Doubleday.

Vogel, Arthur L. 1937. *A Text-book of Qualitative Chemical Analysis.* London: Longman, Green and Co.

+ Vogel, Arthur I., and G. Svehla. 1996. *Vogel's Qualitative Inorganic Analysis,* revised. Harlow, England: Longman.

Walker, Jearl. 1981. The Amateur Scientist. *Scientific American* **244**(2): 168–75.

White, Raymond. 1984. The Characterization of Proteinaceous Binders in Art Objects. *Technical Bulletin* **8**: 5–14. London: National Gallery.

+ Wiig, Edwin O., Willard R. Line, and John F. Flagg. 1954. *Semimicro Qualitative Analysis.* New York: D. Van Nostrand.

+ Williams, R. Scott, 1989. The Beilstein Test: A Simple Test to Screen Organic and Polymeric Materials for the Presence of Chlorine. *CCI Notes* No.17/1 Ottawa: Canadian Conservation Institute.

+ Williams, R. Scott, 1989. The Diphenylamine Spot Test for Cellulose Nitrate in Museum Objects. *CCI Notes* No. 17/2 . Ottawa: Canadian Conservation Institute.

Windholz, Martha (ed.) 1983. *The Merck Index, An Encylopedia of Chemicals, Drugs, and Biologicals.* 10th edn. Rahway, NJ: Merck & Co.

Zhang, Jinping, David Thickett, and Lorna Green. 1994. Two Tests for Detection of Volatile Organic Acids and Formaldehyde. *Journal of the American Institute for Conservation* **33**(1): 47–53.